ABOUT ISLAND PRESS

Island Press, a nonprofit organization, publishes, markets, and distributes the most advanced thinking on the conservation of our natural resources—books about soil, land, water, forests, wildlife, and hazardous and toxic wastes. These books are practical tools used by public officials, business and industry leaders, natural resource managers, and concerned citizens working to solve both local and global resource problems.

Founded in 1978, Island Press reorganized in 1984 to meet the increasing demand for substantive books on all resource-related issues. Island Press publishes and distributes under its own imprint and offers these services to other nonprofit organizations.

Support for Island Press is provided by Apple Computers, Inc., Mary Reynolds Babcock Foundation, Geraldine R. Dodge Foundation, The Charles Engelhard Foundation, The Ford Foundation, Glen Eagles Foundation, The George Gund Foundation, The William and Flora Hewlett Foundation, The Joyce Foundation, The John D. and Catherine T. MacArthur Foundation, The Andrew W. Mellon Foundation, The Joyce Mertz-Gilmore Foundation, The New-Land Foundation, The J. N. Pew, Jr., Charitable Trust, Alida Rockefeller, The Rockefeller Brothers Fund, The Florence and John Schumann Foundation, The Tides Foundation, and individual donors.

BALANCING
ON THE BRINK
OF EXTINCTION

BALANCING ON THE BRINK OF EXTINCTION

THE ENDANGERED SPECIES ACT
AND LESSONS FOR THE FUTURE

Edited by Kathryn A. Kohm

ISLAND PRESS

Washington, D.C. □ *Covelo, California*

No copyright claim is made in "Gap Analysis of Species Richness and Vegetation Cover: An Integrated Biodiversity Conservation Strategy," a work produced by employees of the U.S. government.

Seven articles of this text were published previously in a special issue of *Endangered Species Update*, a publication of the School of Natural Resources at the University of Michigan.

Grateful acknowledgment is expressed for permission to reprint "The Endangered Species Act: Legislative Perspectives on a Living Law," which appeared originally as a foreword to *The Endangered Species Act: A Guide to Its Protections and Implementation*, published by the Stanford Environmental Law Society. Copyright, Trustees of Stanford University.

Grateful acknowledgment to Julie Kierstead for the use of endangered species plant illustrations.

Library of Congress Cataloging-in-Publication Data

Balancing on the brink of extinction : the Endangered Species Act and lessons for the future / edited by Kathryn A. Kohm.
p.   cm.
Includes index.
ISBN 1-55963-007-8.—ISBN 1-55963-006-X (pbk.)
1. Endangered species—Law and legislation—United States. I. Kohm, Kathryn A.
RF5640.B35   1991
346.7304'695—dc20
[347.3064695]   90-5031
                                                                                                            CIP

Printed on recycled, acid-free paper

Manufactured in the United States of America

10   9   8   7   6   5   4   3   2   1

ACKNOWLEDGMENTS

Translating this book from an intriguing idea into reality could not have been possible without the support and inspiration of many people. I would like to thank all of the authors who found time in their already overcrowded schedules to contribute their ideas, insights, and expertise. They patiently stuck with the project through many drafts and revised timelines, and their time and energy has been much appreciated.

I am also grateful to Barbara Dean at Island Press. She supported this project from the beginning, and her advice and expertise truly made this book a reality.

Chuck Ayers deserves special thanks not only for his careful reading of many of the essays, but also for his untiring encouragement and patience that kept me going throughout the project.

The University of Michigan's School of Natural Resources provided organizational support for the production of the special issue of *Endangered Species Update* from which this book evolved.

Finally, I would like to thank Michael Soulé. He reviewed several of the essays for this volume and, as always, offered constructive criticism that pushed me to improve my work. Most important, he served as a model of dedication to the cause of species conservation, excellent writing, and inspiring thinking. He has been a great mentor, and I dedicate this book to him.

CONTENTS

ACKNOWLEDGMENTS   vii

INTRODUCTION   3
  Kathryn A. Kohm

THE ACT'S HISTORY AND FRAMEWORK   10
  Kathryn A. Kohm

PART I
REFLECTIONS ON THE ACT

THE ENDANGERED SPECIES ACT: LEGISLATIVE PERSPECTIVES
  ON A LIVING LAW   25
  Congressman John D. Dingell

THE POWER AND POTENTIAL OF THE ACT   31
  Lynn A. Greenwalt

LOOKING BACK OVER THE FIRST FIFTEEN YEARS   37
  Michael J. Bean

LIFE IN JEOPARDY ON PRIVATE PROPERTY   43
  Holmes Rolston III

ix

SNAIL DARTERS AND PORK BARRELS REVISITED: REFLECTIONS
  ON ENDANGERED SPECIES AND LAND USE IN AMERICA   62
  George Cameron Coggins

PART II
COMPONENTS OF THE FEDERAL ENDANGERED
SPECIES PROGRAM

THE ENDANGERED SPECIES LISTS: CHRONICLES OF EXTINCTION? 77
  William Reffalt

AVOIDING ENDANGERED SPECIES/DEVELOPMENT
  CONFLICTS THROUGH INTERAGENCY CONSULTATION   86
  Steven L. Yaffee

FEDERALISM AND THE ACT   98
  John P. Ernst

INTERNATIONAL IMPLEMENTATION: THE LONGEST ARM OF
  THE LAW?   114
  Mark C. Trexler and Laura H. Kosloff

THE APPROPRIATIONS HISTORY   134
  Faith Campbell

IMPLEMENTING RECOVERY POLICY: LEARNING AS WE GO?   147
  Tim Clark and Ann Harvey

PART III
IMPLEMENTATION CHALLENGES

WESTERN WATER RIGHTS AND THE ACT   167
  A. Dan Tarlock

INVERTEBRATE CONSERVATION   181
  Dennis D. Murphy

PREDATOR CONSERVATION   199
  Kevin Bixby

PESTICIDE REGULATION   214
  Jim Serfis

PART IV
CONSERVING BIODIVERSITY

FROM ENDANGERED SPECIES TO BIODIVERSITY 227
    Reed F. Noss

IN SEARCH OF AN ECOSYSTEM APPROACH TO ENDANGERED
    SPECIES CONSERVATION 247
    Hal Salwasser

COPING WITH IGNORANCE: THE COARSE-FILTER STRATEGY
    FOR MAINTAINING BIODIVERSITY 266
    Malcolm L. Hunter, Jr.

GAP ANALYSIS OF SPECIES RICHNESS AND VEGETATION
    COVER: AN INTEGRATED BIODIVERSITY CONSERVATION
    STRATEGY 282
    J. Michael Scott, Blair Csuti, Kent Smith, J. E. Estes,
    and Steve Caicco

NEEDED: AN ENDANGERED HUMANITY ACT? 298
    Anne H. Ehrlich and Paul R. Ehrlich

INDEX 303

ABOUT THE CONTRIBUTORS 313

BALANCING
ON THE BRINK
OF EXTINCTION

INTRODUCTION

by

KATHRYN A. KOHM

IN APRIL 1978, almost fifteen years after the Tennessee Valley
Authority (TVA) proposed construction of a multimillion-dollar
dam on the Little Tennessee River, one of the most controversial
endangered species protection cases was heard in the nation's
highest court.[1] In posing his challenge to the Endangered Species Act, Griffin Bell, the counsel for TVA, pulled a small glass jar
from his briefcase. The jar contained one tiny snail darter. As
Bell held it up before the Supreme Court, he asked how a three-
inch fish could be valued higher than a multimillion-dollar dam.
Clearly the benefits of the Tellico Dam and the price of the snail
darter's existence as a species do not fit neatly into a cost-benefit
analysis. After all, we are taught in the most basic math class
that one cannot compare apples and oranges in the same equation. Yet the image created by Griffin Bell epitomizes precisely
this dilemma. Indeed it is the dilemma that the Endangered
Species Act of 1973 requires us to confront. The act is one of
those extraordinary collective decisions that emerge in a pluralist democracy in which innovation and progress are often pain-
fully slow. It requires us to look beyond the day-to-day business
of building dams, harvesting timber, or improving crop yields
and learn to live in a way that is compatible with our fellow
species.

During the same years that the Endangered Species Act was
being forged, men walked on the moon. Although the two events

3

will probably not go down in the history books together, there is a common thread that marks them as significant steps forward in our collective understanding of ourselves and the earth. Both events forced us to consider ourselves from a new vantage point. Those first NASA photographs of our tiny planet suspended in the blackness of space confronted us with a planetary perspective of our size and fragility in the cosmos. The Endangered Species Act confronts us with an ecological perspective of ourselves in the biosphere. As Senator Patrick Leahy stated during the debate over the 1978 amendments to the act: "Ultimately, we are the endangered species. *Homo sapiens* are perceived to stand at the top of the pyramid of life, but the pinnacle is a precarious station. We need a large measure of self-consciousness to constantly remind us of the commanding role which we enjoy only at the favor of the web of life that sustains us, that forms a foundation of our total environment. We share the planetary gene pool with that snail darter in the Little Tennessee River."[2]

The more we learn about the earth's natural history, the more we have come to realize that we are in the midst of an extinction episode of historic proportion. There have been other extinction episodes in the history of life on earth.[3] At the end of the Cretaceous period, a large number of marine and terrestrial species disappear from the fossil record. Among them are the dinosaurs. Apparently all evolutionary lineages of the dinosaurs went extinct approximately 65 million years ago. Further back in geological time, the fossil record shows an even more devastating extinction spasm during the late Permian (250 million years ago). During that time, the fossil record suggests that as many as 52 percent of all existing families of marine animals were lost. Between these "mass extinction episodes," it appears there were also a number of smaller extinction events that scientists often use to mark the epochs and periods of geological time. Overall, it is estimated that the current stock of species represents between 2 and 10 percent of all species that have ever lived on earth.[4] In other words, extinction is a biological reality. What sets the current extinction episode apart from the others, however, is that it is largely being caused at the hands of a single species. Moreover, the current rates of species loss appear to be unprecedented. Evidence in the fossil record suggests that historical extinction rates are trivial compared to those of the past several decades. Norman Meyers has suggested that by the end of this century, we could lose as many as a hundred species per day due to habitat destruction and other assaults on the natural world.[5] This is to be compared to an estimated background level of "natural" extinction averaging a few species per million years for most kinds of organisms.[6] Clearly it is extremely difficult to estimate extinction rates to a high level of accuracy, particularly in light of the fact that we have identified only a fraction of the species believed to exist. Nevertheless, amidst the figures and calculations, at least one poignant conclusion stands out: We are balancing on the brink of causing the largest extinction episode in 65 million years.

With at least a partial understanding of the magnitude of the extinction problem, Congress passed the first Endangered Species Preservation Act in 1966. The 1966 act was followed by two updated versions, one in 1969 and the other in 1973. The 1973 version has now survived over one and a half decades of American politics. Although there have been times when it appeared as though the rug was about to be pulled out from beneath it, the basic framework of the act has remained intact. The history of the first seventeen years of this truly extraordinary piece of legislation is the subject of this book.

The book evolved from a special issue of *Endangered Species Update*, a publication of the School of Natural Resources at the University of Michigan. The original eight articles in the special issue are presented here along with fourteen additional essays written by a diverse group of authors working on species preservation. The intent is to pose, explore, and begin to answer two fundamental questions: What have we learned about endangered species protection since the passage of the 1973 act? And based on our experience thus far, where should we direct future conservation efforts? To borrow a line from David Ehrenfeld's essay in *The Last Extinction*, "The future is shy. If you want to catch a glimpse of it, you have to sneak up on it from behind. So the place to start for a look into the future is the past."[7] This notion underlies the character and design of the book. Ultimately the purpose of a good retrospective should be to gain better perspective.

The essays are divided into four parts. The essays in Part I take an overarching look at our experience with the federal endangered species program. They discuss the legal, political, and philosophical ramifications of the act. In Part II, each essay focuses on a particular component of the act, reporting on strengths and weaknesses and reflecting on problems and potential remedies. Part III considers four areas where implementation of the act has been particularly challenging: water rights, invertebrate conservation, predator protection, and pesticide regulation. Finally, Part IV takes up a theme reiterated throughout the book: If we are to cope with the ever increasing number of endangered species, we must move from the current species-by-species approach to conservation to a broad strategy of protecting ecosystems and conserving biodiversity. Yet the question of what an "ecosystem" approach ought to look like remains open. The essays in Part IV explore these issues as they relate to the future of the federal endangered species program.

Each of the essays is a mix of reporting, reflection, and prescription. Although they are written from different perspectives and in varying styles, several common themes stand out. First, our increased sensitivity to the plight of a relatively small number of species officially listed as threatened or endangered has opened our eyes to the severity of the extinction crisis. At least some of this awareness has been stimulated by the Endangered Species Act; it has certainly played a major role in fostering research, debate, and public concern. Yet one of the harshest lessons of the past two decades is that we grossly underestimated the magnitude of the problem as well as the resources needed to slow the wave of extinction facing us. In the early years when the first federal legislation was being developed, the issue of species extinction was primarily defined as a technical problem requiring technical solutions.[8] "Solving" the endangered species problem was largely viewed as a matter of acquiring habitat and funding captive breeding programs for a select number of species. In his essay in Part I, Lynn Greenwalt recalls that many members of Congress originally "thought they were voting to protect eagles, bears, and whooping cranes and failed to make the connection to questions about irrigation projects, timber harvests, the dredging of ports, or the generation of electricity." Indeed, as William Reffalt points out in Part II, the 1966 version of the Endangered Species Act was passed with the intent of protecting only about thirty-five species of mammals and thirty to forty species of birds believed to be near extinction. There are now over a thousand species on the federal list of threatened and endangered species—and this only begins to scratch the surface of the problem. Some 3,900 species have been identified as candidates for listing in the United States alone. In tropical countries, extinction rates dwarf these figures. This is the backdrop against which any strategy for species conservation must be evaluated.

Yet if the extinction problem is greater than we had ever imagined, so too is the power and scope of the act itself. The act has been criticized for its lack of flexibility. But the absolute mandate, which has proved remarkably resilient, has provided strong incentives for parties involved in species conservation cases to look beyond the courtroom for resolution of conflicts. The initial response of developers, whose plans jeopardize a species' existence, often has been to set up a win/lose battle pitting economic development against species conservation. Yet this sort of adversarial approach, typified by the Tellico Dam case, is increasingly costly—and ultimately self-defeating. In his essay in Part III, Dan Tarlock points out that the judicial decisions favoring species protection in water rights cases have stimulated interest in alternative ways to meet the obligations of the act and still permit water development. In his discussion of pesticide regulations, Jim Serfis reaches a similar conclusion: "Because protection may be achieved in many different ways, there is a tremendous though often untapped potential to devise solutions that protect species while allowing prudent use of certain pesticides."

The Section 7 consultation process offers a structure within which diverse groups might pursue creative problem solving. In Part II, Steven Yaffee elaborates on this idea, suggesting that the endangered species problem is principally a land-use problem and that the consultation process has "gotten the endangered species program into land-use planning through the back door." Similarly, in Part I, George Coggins illustrates how the act "has had broad and deep consequences for land use in America."

Another theme, although not new, is a pivotal issue. To win philosophical support for endangered species protection as an abstract concept is relatively easy; to garner the necessary resources and political support to implement on-the-ground protection programs is another question. In his essay in Part III, Kevin Bixby suggests that the easy part is over for predator conservation: "Taxonomic questions have been resolved, recovery plans have been formulated, and captive breeding programs have been established. The challenge now is to make predator recovery and protection work on the ground." But inadequate resources and the lack of commitment shown by many administering agencies severely weaken our ability to implement the act and provide substantial protection for many species. In Part II, Faith Campbell discusses the problem in her review of the appropriations history of the act. Yet the infusion of more dollars and more personnel, though sorely needed, is not enough. Fulfilling the Endangered Species Act's mandate demands leadership backed by a strong constituency. In Part IV, Hal Salwasser argues that a fundamental ingredient of ecosystem-based conservation strategies is an informed and supportive constituent base. In Part II, Steven Yaffee acknowledges that "endangered species will only be protected so long as there is a political will to do so."

Finally, in the next several decades we must move from the reactive measures that characterize current programs to bold, active strategies. The situation might be likened to a treadmill that is exponentially increasing in speed. As more and more species are pushed to the brink of extinction, our chances of keeping up with the problem by trying to run faster in the same direction will only diminish. In their essay in Part IV, Mike Scott and colleagues write that "ultimately it is easier and more cost-effective to protect intact, functioning ecosystems than to forever race to save individual species in imminent danger of extinction." Our focus, then, must turn from the preservation of individual species toward the protection of the full spectrum of biological diversity including communities, ecosystems, and regional landscapes.

In the face of sweeping technological change—and what appears to be the inevitable loss of enormous numbers of species—it behooves us to move purposively and resolutely in future conservation efforts. Drawing lessons from the past will be critical. I hope that the thoughts expressed in this book constitute the beginning of a dialogue rather than an end in itself. With so much at stake, we can hardly afford to do otherwise.

Notes

To obtain a list of endangered species and a copy of the Endangered Species Act, write to:

U.S. Fish and Wildlife Service
Publications Unit
ARLSQ Room 130
Washington, DC 20240

Copies of recovery plans and certain other scientific documents are available for purchase through:

U.S. Fish and Wildlife Reference Service
5430 Grosvenor Lane
Suite 110
Bethesda, MD 20814
1-800-582-3421
1-301-492-6403 (in Maryland)

1. *TVA v. Hill*, 437 U.S. 153 (1978).

2. Quoted in Norman Meyers, *The Sinking Ark: A New Look at the Problem of Disappearing Species* (Oxford and New York: Pergamon Press, 1979), p. ix.

3. David Raup, "Diversity Crises in the Geological Past," in *Biodiversity*, ed. E.O. Wilson (Washington, D.C.: National Academy Press, 1988), pp. 51–57.

4. Paul Ehrlich and Anne Ehrlich, *Extinction: The Causes and Consequences of the Disappearance of Species* (New York: Random House, 1981).

5. Meyers, *The Sinking Ark*.

6. David Jablonski, "Mass Extinctions: New Answers, New Questions," in *The Last Extinction*, ed. Les Kaufman and Kenneth Mallory (Cambridge: MIT Press, 1986), pp. 43–61.

7. David Ehrenfeld, "Life in the Next Millennium: Who Will Be Left in Earth's Community," in *The Last Extinction*, ed. Les Kaufman and Kenneth Mallory (Cambridge: MIT Press, 1986), pp. 167–186.

8. Steven L. Yaffee, *Prohibitive Policy: Implementing the Federal Endangered Species Act* (Cambridge: MIT Press, 1982).

THE ACT'S HISTORY AND FRAMEWORK

by

KATHRYN A. KOHM

The purposes of this Act are to provide a means whereby the ecosystems upon which endangered species and threatened species depend may be conserved, to provide a program for the conservation of such ... species, and to take such steps as may be appropriate. ...

THE ENDANGERED SPECIES ACT of 1973 fits into a long history of increasing federal involvement in wildlife conservation. Until 1900, jurisdiction over wildlife remained largely with the states where conservation efforts were focused almost exclusively on traditional game species. Near the turn of the century, however, the balance of power over wildlife began to shift. By the late 1800s, well-organized commercial interests had overrun the efforts of individual states to enforce their wildlife laws by killing enormous quantities of wildlife in one state and quickly transporting them to another. In response, Congress passed the Lacey Act of 1900, making a significant—if cautious—entry into the field of wildlife regulation. Using the constitutional power granted in the Commerce Clause, Congress initially passed the Lacey Act to bolster enforcement of existing state wildlife regulations.[1] In addition, the Lacey Act included several provisions for affirmative species conservation on the federal level. Among these was an authorization for the secretary of agriculture to take necessary steps toward the "preservation, distribution, introduction, and restoration of game birds and other wild birds" subject to existing state laws. This marked Congress's first official recognition (albeit on a limited basis) that the loss of species is an issue of national concern.

In the series of federal wildlife laws that followed the Lacey Act, elements of the current federal endangered species program began to emerge (Yaffee, 1982). Federal regulation of the taking of species was first undertaken on federal lands. As early as 1894, hunting was prohibited in Yellowstone National Park; in 1906, hunting of birds on federal lands reserved as breeding grounds for birds was prohibited. Although it was declared unconstitutional by a federal district court, the Migratory Bird Act of 1913 was the first federal statute to attempt to regulate taking more broadly. It prohibited the hunting of all migratory and insectivorous birds except under federal regulations and declared those species to be within federal custody. The idea was revived a few years later, however, with passage of the Migratory Bird Treaty Act of 1918. The 1918 act provided for taking regulations similar to those of the 1913 act and was upheld by the Supreme Court in 1920.[2]

Habitat acquisition began as early as 1903 with the designation of Pelican Island National Wildlife Refuge. Over the following two decades, both Congress and the president established refuges throughout the country. By 1929, Congress had established a commission through the Migratory Bird Conservation Act to review Interior Department proposals for refuge purchases. Dependable funding for refuge acquisitions was later provided by the 1934 Migratory Bird Hunting Stamp Act and the Pittman-Robertson Act. Finally, the Fish and Wildlife Coordination Act of 1934 set a precedent for requiring federal agencies to consider the effect of their actions on wildlife populations and advocated state/federal cooperation to develop a national wildlife conservation program.

The maturation of federal wildlife law was propelled by increasing concern over the loss of species among wildlife professionals and the general public. By the 1960s, a strong constituency for species conservation had developed. This base provided the political climate for the creation of a Committee on Rare and Endangered Species in the Interior Department's Bureau of Sport Fisheries and Wildlife in 1964. The committee, composed of nine biologists, published the "Redbook"—the first official endangered species list. That list (based solely on informal expert opinion) cited sixty-three vertebrate species thought to be in danger of extinction. The Redbook, along with increased media attention to the plight of many species, reflected a growing tide of environmental awareness—and in turn endangered species became a powerful symbol of the need for environmental protection.[3] Congress responded with three successive endangered species statutes, each building upon and adding to the conviction and specificity of the others.

THE 1966 ACT

Although limited in scope, specificity, and enforcement capacity, the 1966 Endangered Species Preservation Act paved the way for a comprehensive federal endangered species program. Unlike earlier efforts to provide protection on a species-by-species basis, the authors of the 1966 act sought to outline a more inclusive program for species protection. In a cover letter accompanying the draft legislation, Interior Secretary Stewart Udall stated that "the principal objective of this proposed legislation is to authorize and direct the Secretary of the Interior to initiate and carry out a comprehensive program to conserve, protect, restore, and where necessary to establish wild populations [and] propagate selected species of native fish and wildlife ... that are found to be threatened with extinction."[4]

The actual provisions of the act, however, were not as far-reaching as Udall had envisioned. For the most part, the act was a vague policy directive that served primarily as a symbolic statement of congressional support for endangered species protection. One of the act's more substantive features was its authorization of acquisition authority and funds to purchase habitat for endangered species. The interior secretary could use up to $15 million ($5 million annually and no more than $750,000 for any one area) from the Land and Water Conservation Fund for such purposes.

The interior secretary was also directed to publish a list of native species threatened with extinction in the *Federal Register*. A species was to be listed if "its habitat is threatened with destruction, drastic modification, or severe curtailment, or because of overexploitation, disease, predation, or because of other factors." In making such a determination, the secretary was required only to seek the advice and recommendations of interested persons and to consult with states in which the species was found. No other procedures were specified.

Interestingly, the 1966 act did reflect recognition of the need for interagency cooperation in the implementation of a federal endangered species program. The act's language in this respect was vague, however, and unenforceable. The interior secretary was required to use programs within his jurisdiction to further the purposes of the act, and to encourage other agencies to do the same, but only "to the extent practicable." Similarly, the act called upon all agency heads to protect endangered species of native fish and wildlife and, "insofar as practicable and consistent with the primary purposes of such bureaus, agencies, and services, [to] preserve the habitats of such threatened species on lands under their jurisdiction."

Perhaps more telling were the provisions omitted from the 1966 act. With the limited exception of a prohibition on the taking of wildlife on reserves designated specifically for the conservation of an endangered species, no taking restrictions (that is, restrictions on killing, trapping, collecting, and harming individuals of an endangered species) were included for endangered species. No restrictions were placed on interstate commerce. Moreover, the act implied coverage only of native species, excluding even subspecies. Nevertheless, the act was hailed as a great achievement. Perceived largely as a "no-lose" bill (having no significant effect on commercial activity), it won an easy victory in Congress. As such it put species protection on the congressional agenda and set the stage for subsequent statutes.

THE 1969 ACT

The 1969 Endangered Species Protection Act incrementally began to resolve some of the ambiguities of the earlier statute. On the domestic front, the 1969 act extended the Lacey Act's ban on interstate commerce to include reptiles, amphibians, mollusks, and crustaceans. The inclusion of reptiles and amphibians was, in large part, a response to uncontrolled alligator poaching in the Southeast. Furthermore, the term "fish and wildlife" was explicitly defined to include vertebrate as well as invertebrate species. Acquisition authority was also expanded, though only marginally. Funding up to $1 million was provided annually over a three-year period for the purchase of private holdings within areas already managed by the Department of Interior; further, the amount that could be spent on a single purchase was increased from $750,000 to $2.5 million.

The more profound changes, however, were made in relation to international wildlife conservation. For the first time, Congress officially recognized the international dimension of the extinction crisis. The 1969 act directed the interior secretary to promulgate a list of species "threatened with worldwide extinction" and to prohibit importation of these species, except in certain cases. The exceptions were carved out as a result of the first significant political conflict encountered by proponents of the endangered species program. Realizing the commercial implications of a unilateral import ban, the U.S. fur industry launched a lobbying campaign against the ban. In the end, the secretary was given the power to grant permits for the importation of endangered species or their products for zoological, educational, or scientific purposes or for captive propagation programs. To prevent "undue economic hardship," the secretary could also allow parties that had entered into a contract for the importation of a species (or product) before it was listed to continue their trade for up to one year.

Another important international element of the 1969 act was its directive to the interior secretary to foster a coordinated international effort for endangered species conservation. The act specifically called for an international conference in order to reach a binding convention on the conservation of endangered species. This meeting, designed as an answer to commercial objections to unilateral action, was to be convened by the secretary by June 30, 1971. Although the original deadline was missed by over a year, that meeting ultimately produced the Convention on International Trade in Endangered Species (CITES). This convention sets up a structure whereby trade restrictions are based on a species' vulnerability to extinction.[5]

Though the actual terms of the convention fell short of those envisioned in the 1969 act, CITES has been one of the most lasting international elements of the U.S. endangered species program.

THE 1973 ACT

In an environmental message to the nation on February 8, 1972, President Nixon declared that current legislation did not provide the management tools needed to save a vanishing species. He then suggested the need for legislation that would make it a federal offense to take any endangered species and would provide protection *before* a species was in critical danger of extinction.[6] His statements came at the height of the environmental movement. They reflected a groundswell of public support for environmental preservation in general and endangered species protection in particular.

This wave of environmental awareness set in motion an array of environmental legislation that included the Endangered Species Act of 1973. The authors of the 1973 act set out not only to remedy the shortcomings of previous acts but also to make a bold collective statement of moral and legal conviction regarding endangered species. Among the act's provisions is a broad prohibition of the taking of endangered species anywhere in the United States; a requirement that no federal agency may jeopardize the continued existence of an endangered species; an extension of protection not only to species presently threatened with extinction but also to those found likely within the foreseeable future to become threatened; and the elimination of limits on land acquisition funds that could be used to purchase endangered species habitat.

Unlike the 1969 act, the 1973 statute was passed with virtually no opposition. It was viewed largely as a symbolic issue with few obvious costs. Even in subsequent reauthorization efforts when it had become clear that endangered species protection meant significant restrictions on commerce and development, support for the basic tenets of the act did not waver. In short, it is difficult (at least in an abstract sense) to argue in favor of extinction. Hence, debate over the 1973 bill centered more on technical and jurisdictional issues than on the comprehensive mandate it represented. Five years later, when the Supreme Court ruled in favor of the snail darter in *TVA v. Hill*,[7] the majority opinion stated that "the plain language of the Act, buttressed by its legislative history, shows clearly that Congress viewed the value of endangered species as incalculable."

Listing of Species

The authority to list species as threatened or endangered is shared by the National Marine Fisheries Service (NMFS), which is responsible for the listing of most marine species, and the Fish and Wildlife Service (FWS), which administers the listing of all other plants and animals. In contrast to previous statutes, the 1973 act provides two classifications under which a species may be listed. Species determined to be in imminent danger of extinction throughout all or a significant portion of their range are listed as "endangered." Species determined likely to become endangered in the foreseeable future are listed as "threatened." Further, distinct populations may be listed even if the species is abundant in other portions of its range. The criteria for endangerment must be based solely on biological evidence and the best scientific and/or commercial data available. Moreover, additions or deletions may be proposed by anyone who presents adequate evidence of the endangered status of a species.

The 1978 amendments to the act substantially altered the original listing process by requiring that each species listing be accompanied by the listing of critical habitat and that economic factors be considered in designating critical habitat. As a practical matter, however, these changes, in addition to numerous new requirements for hearings and local notice, were burdensome and often extremely difficult to carry out. In effect, the entire listing program was stifled. During the first year of the Reagan administration, not a single species was added to the list. In 1982, therefore, Congress relaxed the requirements for listing and attempted to expedite the process by requiring the interior secretary to issue a preliminary finding within ninety days of receiving a listing petition. The 1982 amendments also allowed species listings to proceed prior to the listing of their critical habitat.

Despite the alleviation of certain burdensome listing requirements, an enormous backlog of species has developed in the listing process. By 1988, more than 3,000 species had become stuck in the listing pipeline.[8] To prevent candidate species from further population reductions or even extinction, Congress once again amended the listing program in 1988 to require the interior secretary to monitor the status of candidate species and, if necessary, make use of emergency listing powers.

Interagency Consultation

One of the most far-reaching provisions of the Endangered Species Act is its requirement that federal agencies and departments through their own actions or actions funded or permitted by them must not jeopardize the continued existence of a listed species or its habitat.[9] To fulfill their duties under Section 7, agencies planning actions in an area used by an endangered or threatened species must consult with either the Fish and Wildlife Service or the National Marine Fisheries Service.

The underlying power of this provision was not fully recognized until the landmark decision in *TVA v. Hill*. Plaintiffs sought to stop construction of a multimillion-dollar dam proposed by Tennessee Valley Authority by charging that it jeopardized the last remaining habitat of the endangered snail darter. The case went all the way to the Supreme Court, which ruled that the act, as written, holds endangered species protection above all other considerations. That decision prompted Congress to take a second look at Section 7 in 1978. The basic mandate of Section 7, however, survived the political onslaught. A compromise was forged whereby a committee (nicknamed the "God Committee") was established to review applications for exemptions to Section 7. To grant such an exemption, the committee must determine that no "reasonable and prudent" alternatives exist to the agency's proposed activity and that the benefits of the action clearly outweigh preservation of the endangered species it threatens. To date, few agencies have opted to put a project to this test.

In addition to creating the exemption process, the 1978 amendments formalized the consultation process. In response to the TVA controversy, Congress also prohibited the "irreversible" or "irretrievable" commitment of agency resources to a project until the consultation is completed. For all formal consultations, the FWS or the NMFS must provide a written biological opinion detailing the finding upon which it based its decision. And all consultations must be based on the best scientific and commercial data available.

The Taking Prohibition

The 1973 law declares that it is unlawful to "take" an endangered or threatened species within the United States, its territorial waters, or on the high seas. The act defines the term "take" in the broadest terms to include "harass, harm, pursue, hunt, shoot, wound, kill, trap, capture, or collect, or to attempt to engage in any such conduct." In *Palila v. Hawaii Department of Land and Natural Resources*, the scope of these taking prohibitions was first revealed.[10] The Sierra Club charged that the state of Hawaii had violated the taking prohibition by maintaining a population of feral sheep and goats in the native forests that the endangered palila (a Hawaiian finch) depended upon for nesting sites. In their finding that the destruction of native forest constituted a taking under the 1973 act, the court reasserted the comprehensive nature of the congressional prohibition. The Fish and Wildlife Service has subsequently revised its definition of the word "harm" to read as follows: " 'Harm' in the definition of 'take' in the Act means an act which actually kills or injures wildlife. Such act may include significant habitat modification or degradation where it actually kills or injures wildlife by significantly impairing essential behavior patterns, including breeding, feeding or sheltering."[11] The practical effect of this revision, however, has been minimal. The new definition essentially clarified what the *Palila* court understood the Fish and Wildlife Service's original implementing regulations to require.

In the years since the act was first passed, several exceptions to the taking provision have been granted. Among these are provisions for "experimental populations" and "incidental takings." In 1982, Congress authorized the listing of certain populations of endangered species to be listed as experimental and provided for relaxed taking restrictions for these populations.

An experimental population is one established outside the current range of a species as part of an approved recovery program. The notion behind a relaxed taking prohibition for these new populations is that it gives FWS personnel the flexibility necessary to gain the cooperation of neighboring landowners and other government agencies. For example, problem wolves that are reintroduced into an area and that repeatedly prey on livestock may be taken provided that the survival of the species and the overall recovery program is not jeopardized. Similarly, individuals can apply to the interior secretary for permission to engage in "incidental takings" under limited circumstances. An incidental take is defined as "any taking otherwise prohibited . . . if such taking is incidental to, and not the purpose of, the carrying out of an otherwise unlawful activity." As part of the permit process, applicants must submit a mitigation plan designed to conserve the species in question. If the secretary approves this plan and finds that the taking will not appreciably reduce the likelihood of the survival and recovery of the species in the wild, he may issue an incidental taking permit. As with experimental populations, the underlying idea of this provision is to allow for some management flexibility while maintaining the purpose and conviction of the act.

Until recently, the absolute restrictions on taking endangered or threatened species did not apply to plants. In the 1988 amendments, Congress added to the 1982 amendments prohibiting private collecting of endangered plants on federal lands by further prohibiting activities that maliciously damage listed plants on federal land or harm them on nonfederal land in violation of state laws.

Restrictions on Trade

In addition to the taking prohibition, Section 9 imposes a blanket prohibition on the import or export of an endangered species or any products made from an endangered species. Similarly, the act makes it unlawful to engage in commerce in endangered species or their products or even to possess a listed species taken in violation of the act.

Section 8 provides the implementation authority for CITES. It

also directs the interior secretary to encourage foreign governments to establish endangered species programs. Toward this
end, the secretary is authorized to provide both financial and
technical assistance in support of foreign fish, wildlife, and plant
conservation programs.[12] Finally, Section 8 gives the secretary
authority to implement the Convention on Nature Protection
and Wildlife Preservation in the Western Hemisphere of 1940.
Prior to 1982, the secretary was only authorized to represent the
United States in matters of the convention.

Cooperative Agreements with States

Section 6 sets out procedures for the secretary of agriculture to
enter into cooperative agreements with individual states and
provides funding for their implementation.[13] When a state program for endangered species conservation is determined to meet
a list of criteria specified in Section 6(c), the interior secretary is
required to enter into a cooperative agreement. Once an agreement is signed, the state is eligible for financial assistance up to
75 percent of the cost of the program or as much as 90 percent if
two or more states are involved in a joint program.

Habitat Acquisition

Section 5 provides the interior and agriculture secretaries with
the authority and funds to "carry out a program to conserve fish,
wildlife, and plants, including those which are listed as endangered species or threatened species." Funds are made available
pursuant to the Land and Water Conservation Fund Act of 1965.
The 1973 act set no limits on the maximum amount that could
be spent for endangered species habitat.

Penalties

For those who knowingly violate the act's prohibitions, criminal
penalties of up to $50,000 in fines and one year in prison may be
imposed. In addition the act allows civil damages to be levied up
to $25,000 for each violation. Furthermore, the act contains a

forfeiture provision that allows not only the illegally taken species or products to be seized but also the guns, vessels, vehicles, aircraft, or other equipment used in the taking. And, finally, the Endangered Species Act contains a broad citizen suit provision.

Implementing the provisions of the Endangered Species Act has been a mix of science, politics, and art. And, not surprisingly, there have been considerable discrepancies between theory and practice as the following essays demonstrate. Nevertheless, the basic provisions of the act have proved to be remarkably resilient over the first decade and a half of their existence. The challenge for the future will be to resist attempts to compromise the act's high standards—and, more important, to learn to creatively implement the vision set forth by its authors.

Notes

1. For a detailed discussion of the provisions of the Lacey Act as well as an excellent history of federal wildlife law, see Bean (1983).
2. *Missouri v. Holland*, 252 U.S. 416 (1920).
3. For a detailed discussion of the development of a strong constituency for endangered species protection and the politics behind passage of the Endangered Species Act, see Yaffee (1982).
4. Letter from Secretary of the Interior Steward Udall to Speaker of the House John McCormack, June 5, 1965, reprinted in U.S. House Committee on Merchant Marine and Fisheries, Protection of Endangered Species of Fish and Wildlife, Report No. 1168, 89th Cong., 1st Sess. (Washington, D.C.: Government Printing Office, 1965), pp. 12–14, cited in Yaffee (1982).
5. For further discussion of CITES see the essay by Mark Trexler and Laura Kosloff in Part II of this volume.
6. Weekly Compilation of Presidential Documents 8 (Feb. 8, 1972), pp. 218–224, cited in Yaffee (1982).
7. *TVA v. Hill*, 437 U.S. 187.
8. For further discussion of the listing process for threatened and endangered species see William Reffalt's essay in Part II of this volume.
9. For further discussion of the Section 7 consultation process see Steven Yaffee's essay in Part II of this volume.
10. 471 F. Supp. 985 (D. Hawaii 1979), aff'd, 639 F. 2nd 495 (1981).

11. 46 Fed. Reg. 54748, 54750 (Nov. 4, 1981).
12. For further discussion of the international components of the ESA see the essay by Mark Trexler and Laura Kosloff in Part II of this volume.
13. For further discussion of federal/state cooperation and the Endangered Species Act see John Ernst's essay in Part II of this volume.

References

Bean, Michael J. 1983. *The Evolution of National Wildlife Law.* New York: Praeger.

Coggins, George C. 1975. "Conserving Wildlife Resources: An Overview of the Endangered Species Act of 1973." *North Dakota Law Review* 51:315–337.

Palmer, William D. 1975. "Endangered Species Protection: A History of Congressional Action." *Environmental Affairs* 4:255–293.

Rohlf, Daniel J. 1989. *The Endangered Species Act: A Guide to Its Protections and Implementation.* Stanford, Calif.: Stanford Environmental Law Society.

Yaffee, Steven L. 1982. *Prohibitive Policy: Implementing the Federal Endangered Species Act.* Cambridge: MIT Press.

Statutes Cited

Endangered Species Preservation Act of 1966 (Pub. L. 89-669, 80 Stat. 926, Oct. 15, 1966)

Endangered Species Protection Act of 1969 (Pub. L. 91-135, 83 Stat. 275, Dec. 5, 1969)

Endangered Species Act of 1973 (Pub. L. 93-205, 81 Stat. 884, Dec. 28, 1973; current version at 16 U.S.C. §§1531–1543)

Endangered Species Act Amendments of 1978 (Pub. L. 95-632, 92 Stat. 3751, Nov. 10, 1978)

Endangered Species Act Amendments of 1982 (Pub. L. 97-304, 96 Stat. 1411, Oct. 13, 1982)

Endangered Species Act Amendments of 1988 (Pub. L. 100-478, 102 Stat. 2306, 1988)

PART I

REFLECTIONS ON THE ACT

In the relations of man with the animals, with the flowers, with the objects of creation, there is a great ethic, scarcely perceived as yet, which will at length break forth into light and which will be the corollary and complement to human ethics.

—Victor Hugo

The first prerequisite of intelligent tinkering is to save all the pieces.

—Aldo Leopold

THE ENDANGERED SPECIES ACT: LEGISLATIVE PERSPECTIVES ON A LIVING LAW

by

CONGRESSMAN JOHN D. DINGELL

SEVENTEEN YEARS AGO, the Endangered Species Act of 1973 (ESA) became law. Its goal was unparalleled in all history. Our country resolved to put an end to the decades—indeed, centuries—of neglect that had resulted in the extinction of the passenger pigeon and the Carolina parakeet, as well as the near extinction of the bison and many other species with which we share this great land. If it were possible to avoid causing the extinction of another species, we resolved to do exactly that. It was my distinct pleasure to serve then as chairman of the Subcommittee on Fisheries and Wildlife Conservation and Environment of the House Committee on Merchant Marine and Fisheries and to introduce the bill, H.R. 37, which eventually became the Endangered Species Act.

When Congress passed the Endangered Species Act, it set a clear public policy that we would not be indifferent to the destruction of nature's bounty. Our duty to stem that destruction derives from more than ethical considerations, though such considerations would be a sufficient basis for action. Living plants and animals have, through the centuries, developed a means of coping with disease, drought, predation, and myriad other threats. Understanding how they do so enables us to improve the pest and drought resistance of our crops, discover new medicines for the conquest of disease, and make other advances vital to our welfare. Living wild species are like a library of books still unread. Our heedless destruction of them is akin to burning that library without ever having read its books. The Endangered Species Act is the means by which we seek to avoid complicity in that senseless destruction.

Some people in 1973, and unfortunately still some today, belittle the goals of this great act by belittling the species it seeks to protect. How easy it is to dismiss the protection of a fish, a mollusk, even a plant, as a frivolity, an example of foolish environmental excess. But who will belittle the lowly mold from which the wonder drug penicillin was discovered? Who will belittle the rosy periwinkle, a species of African violet? Had it been allowed to become extinct, we would be without the drug that has made it possible for most victims of childhood leukemia to survive that dread disease. Preventing the extinction of our fellow creatures is neither frivolous nor foolish. It is the means by which we keep intact the great storehouse of natural treasures that make the progress of medicine, agriculture, science, and human life itself possible.

The Endangered Species Act, like our United States Constitution, was written as a flexible document, but durable enough to withstand the evolutionary alterations that have since occurred. Yet in 1978 the ESA, belittled and nearly eradicated, withstood harsh attacks through negotiations between environmentalists and industry. The construction of the Tellico Dam project in the Tennessee Valley Authority (TVA) system was near completion when in 1977 it was halted because of a rare fish called the snail darter. Before its construction, the dam had repeatedly been the subject of attack from the area's property owners concerned about the impact of the dam on property values and actual land possession. The 1973 discovery of the rare snail darter in waters of the Tellico project, and its subsequent listing as an endangered species, resulted in litigation to stop construction. In January 1977, the Sixth Circuit U.S. Court of Appeals upheld the operation of the act and halted dam construction. The Tellico developers, outraged that their project could be ended because of a conflict with an endangered species, appealed the case to the United States Supreme Court. After much deliberation, the court upheld the lower court decision in June 1978. The Supreme Court's ruling sparked a major assault on the very heart of the Endangered Species Act.

Spearheaded by congressional leaders from the Tennessee delegation, the industry sought to eliminate Section 7 of the act, which requires federal agencies to take the necessary steps to prevent any destruction of habitat of an endangered species. The elimination of Section 7 would have gutted the effective protective mechanism of the act. Fortunately, a compromise was reached to create an independent board and an administrative procedure to resolve conflicts between federal projects and endangered species habitat. The first decision by the board was the Tellico Dam project and, once again, the project was put to rest. But not forever.

In 1979, over my objections and those of others intent on preserving the integrity of the ESA, Congress passed a measure to continue construction of the Tellico Dam project. The dam's proponents made an end run around the act by attaching a rider to a House energy and water appropriations bill. Despite considerable opposition to this legislatively created exception to the act, the Senate narrowly voted to continue the project after the snail darters were removed to nearby waters.

In fifteen years, we have learned it is possible to reverse the road to extinction. In 1973, the symbol of our nation, the bald eagle, was en route to extinction because the pesticide DDT had so poisoned its environment that the eagle could no longer lay hatchable eggs. When the government proposed to ban DDT, cotton farmers, citrus growers, and countless others rushed to tell us they could not possibly stay in business without DDT. The sky did not fall, we still have strong cotton and citrus industries, and the bald eagle is well on the road to recovery, aided by active programs of protection and restoration under the authority of the Endangered Species Act.

Other travelers on the road to extinction have turned around and begun the road to recovery. More whooping cranes fly south across the U.S.–Canadian border each fall and return north each spring than at any time in the past half century. Added insurance for the survival of that species has been purchased in the form of a new population established by the ingenious method of putting whooper eggs into the nests of sandhill cranes. Similar intensive management efforts aided by the Endangered Species Act have made possible the reintroduction of the peregrine falcon into the eastern United States, from which it had once been completely extirpated. The American alligator, once decimated throughout the South by poachers supplying illegal leather markets, has rebounded dramatically. It is no longer classified as endangered. Neither is the black pelican in the Southeast, where pesticide poisoning once drove it to the brink of extinction.

Behind these success stories and others like them lies a truth too seldom recognized. From fifteen years of experience with the Endangered Species Act, we have learned that it is almost always possible to conserve endangered species—and thereby promote our long-term welfare—without significantly harming our short-term interests. The number of truly irreconcilable conflicts between endangered species and worthy development projects is astonishingly small. So too is it possible to adjust the ways in which we do business to benefit endangered species without harming our business. The recent effort to reduce the drowning of endangered sea turtles in the shrimp fishery is no exception. Turtle-excluder devices, four or five varieties of which have been developed by fishermen themselves, offer a means of giving essential protection to several imperiled species without harming the shrimp industry. They save turtles, and they catch shrimp. They are a positive solution to a serious environmental problem, a solution that can benefit both the environment and the shrimp industry. All that is needed is the will to make the transition to their use, just as our farmers made the transition from DDT to less hazardous pesticides not so many years ago. Farmers who had used DDT all their lives were understandably reluctant to give it up when its hazards became known. But once they did, the miraculous recovery of the bald eagle, the symbol of the nation, resulted.

Though many great successes have been achieved under the Endangered Species Act, other efforts have ended in disappointment or failure. On June 16, 1987, the last dusky seaside sparrow, a songbird of Florida's Atlantic coastal marshes, died in captivity. Across the continent, in California, the Palos Verdes blue butterfly has vanished. The California condor and the black-footed ferret, two species that were targets of rescue efforts even before the Endangered Species Act was passed, no longer survive in the wild, though captive populations of each may lead to their eventual reintroduction.

Thus while the Endangered Species Act has enabled us to make great progress in protecting many species, the problem to which it is directed remains very much with us. Each day, somewhere in the world, a desperate drama of survival takes place with little notice and little fanfare, but with vital consequences for our future. Despite our efforts, the world continues to experience an alarming, and accelerating, loss of its wild plant and animal species. Scientists believe that never, in all of human history, has the rate of extinction been so rapid as it is today. Human activity may be wiping out yet another species every day, whereas perhaps only one species a century disappears through natural causes.

The Endangered Species Act commits us to make our very best efforts to stem these unprecedented and irreversible losses. Today, slightly more than a thousand species enjoy the nominal protection of the Endangered Species Act. Somewhat more than half of these occur in the United States and its territories; the remainder are found entirely outside U.S. borders. The number of species from within the United States that have been identified as deserving the act's protection, but are not yet listed for protection, exceeds a thousand—twice the number of species currently listed. Most of the species in danger of extinction will wait years, perhaps decades, before they are listed and protected under the act. In a time of deficit reduction, Congress simply has not made available the resources necessary even to list these species formally for protection, much less carry out the action necessary to ensure their survival and recovery. For species already listed, the nominal protection afforded by that listing may be all that the species receives. Recovery plans have thus far been prepared for less than half the listed species, and most of these plans are yet to be implemented.

In the years since the Endangered Species Act became law, we have learned that the problem it seeks to solve is far more serious and affects far more species than previously understood. At the same time, however, we have also learned that the reasons for preventing the extinction of other species are even more compelling. The discovery, within that short period, of the principles of biotechnology have added an urgent new reason to protect the genetic diversity of nature. Suddenly we have learned how to harness the unique genetic attributes of one organism and implant them in another. This discovery has opened the possibility for advances in agriculture and medicine undreamed of two decades ago. But the raw material for this potential revolution in human welfare, bestowed upon us by nature's great diversity, is threatened by our mindless destruction of its diversity. Herman Melville's classic observation that "there is no folly of the beasts of the earth which is not infinitely outdone by the madness of men" aptly describes our willingness to countenance the destruction of yet another species, while its very peculiarity may hold the key to advances in human welfare.

To fulfill the commitment Congress made in 1973, we need to reinvigorate the nation's endangered species program. In particular, we need to provide it with the resources necessary to carry out its basic objectives. The administration that assumed office in January 1989 could make no clearer signal of its commitment to an improved environment than by seeking the expanded resources necessary to carry out the Endangered Species Act.

THE POWER AND POTENTIAL
OF THE ACT

by

LYNN A. GREENWALT

THE ENDANGERED SPECIES ACT (ESA) became law in December 1973; I had been named director of the U.S. Fish and Wildlife Service two months before. Thus, this remarkable legislation and I have had a long association. This retrospective assessment of one of the most powerful environmental laws of the century is that of a nonlawyer who, with his associates, attempted to realize the potential of the legislation without letting its inherent power become its downfall.

From its beginning, the act was regarded as special. A professor friend of mine from a western university once told me he always asked one of his government classes to characterize the act's fundamental nature. The predominant theme of their responses, he reported, was that the act is in a way theological. It reflects a nation's collective commitment to prevent any species of plant or animal from disappearing—a pledge that transcends the social, economic, and national security issues usually addressed in legislation. When one considers that the act protects the Socorro isopod—a small crustacean that has remained unchanged for millions of years and is now restricted to a few small springs in Mexico and the American Southwest—in the same way that it protects the bald eagle, then it is evident we have made a commitment unlike any other.

One of the first jobs after the act was signed was to devise

31

regulations by means of which the act would be executed. A new body of regulatory direction had to be developed, a chore that took many months and through which the real strength of the act was revealed. Gradually it became evident that the ordinary course of federal business would never be the same as the intentions of the act and the reality of the regulations were felt on a day-to-day basis.

The real strength of the act, however, did not become evident until the tiny snail darter illuminated the issue. The Tellico Dam project was well under way when a citizens' group filed a lawsuit to require the TVA to comply with the Endangered Species Act in order to protect the snail darter. Ultimately the suit resulted in a U.S. Supreme Court determination which declared that if a project jeopardized a properly listed species, the project must give way to the organism in jeopardy.

Only a new law could change that. A modification of the ESA was sought immediately, stimulating a rash of proposals and modifications of the statute. During rounds of congressional hearings, many legislators came forward to say they did not know this new act would protect everything; they thought they were voting for legislation to protect eagles, bears, and whooping cranes. They professed not to understand at the time of passage that this law might raise questions about irrigation projects, timber harvests, the dredging of ports, or the generation of electricity. In short, the gap between ideology and actual behavior began to widen.

As the debate raged on, a few recommendations began to emerge. It was finally concluded that the act should be changed to provide for a special review of conflicts like the Tellico Dam issue. An amendment providing for the appointment of a special review committee was passed (including an important proviso that key members could not delegate their responsibilities as reviewers). Such a committee was convened to study the Tellico situation. After examining information related to the snail darter controversy, the committee found the Tellico Dam project without convincing merit and determined that it could not be allowed to eliminate the snail darter. This decision was followed almost immediately, however, by a skillful application of parliamentary procedures in Congress. As a result, the Tellico Dam project was exempted from all provisions of the act.

One of the ironic results of the case was that TVA, spurred on by the pressure of the issue, found room in its budget to develop a way to protect the little-known snail darter. In the process, they found populations of the tiny, reclusive fish in other streams of the area. The event, at least, increased our general understanding of the effects of humans on other creatures and demonstrated how little is really known about the flora and fauna of the United States.

Yet the interposition of a plant or animal between humans and their desire to build, change, or turn a profit became increasingly controversial. It was alleged on one hand that proponents of the act were trying to stop evolution—or at least that part of the process which causes species to disappear. It was sometimes difficult to convince those who held this view that evolution does not really count if it is speeded up by a bulldozer or a chainsaw. Others maintained that environmentalists would want to save the virus causing smallpox or, on a more practical level, would protect a man-killing grizzly bear instead of its potential victims.

Early in the process of applying the act, the question of how to deal with large predators emerged. The grizzly bear was at the center of some of the first controversies. When the bear was finally listed as threatened, its listing was accompanied by a provision allowing for an annual total "man-caused" take, including a legal harvest, protection of livestock, and human self-defense. In the absence of this flexible approach, the last great predatory mammal in the United States might not have been listed at all.

A similar controversy developed around the timber wolf, a remnant population of which existed in northern Minnesota. Those opposed to protecting the wolf insisted that any real increase in wolf numbers would decimate dairy herds in the area and wreak havoc on the deer populations for which the area is well known. Hence a carefully drawn plan for the management of wolves in the area was developed, and state and federal officers were made available to control depredating wolves. The plan has worked well, though it has not allowed for state-managed sport harvest of wolves; lawsuits brought by pro-wolf groups have prevented the exercise of *that* degree of flexibility.

As the act matured, the often-heard expressions of concern

about its intractability, its inconvenience, and its resistance to the application of practical interpretation have been demonstrated to be without merit. For instance, Section 7 of the act requires federal agencies to "consult" with the U.S. Fish and Wildlife Service to determine whether actions proposed by an agency will lead to the "jeopardy" of any listed species. These consultations, more often than not, have resulted in minor adjustments to a proposed action in order to accommodate the needs of a species. Some consultations take months of discussion among experts; most, however, are concluded with a telephone call or exchange of letters.

In recent years, the act's influence has resulted in innovative solutions that have sparked the enthusiasm of everyone involved. A housing development in California that might have eliminated a rare butterfly was changed to accommodate the needs of the insect. The developer has emphasized and exploited his commitment to this idea, and residents of the new development are proud of their co-dwellers.

The strength of the act has been tested many times but remains unwavering. The means by which the act is carried out, however, have been changed—and in most cases ought to have changed. Laws which provide little room for flexibility, or which are applied without an opportunity for the public to understand them and shape their nature, are doomed in the long run. Yet in spite of its strength, the act is vulnerable; its armor is not seamless. It is vulnerable to political intervention and to decisions that are based on political expediency rather than what is best for a species. It is not easy to resist the pressure to make special arrangements to allow the advance of a project. Proponents may justify such accommodations on the grounds of national security, overwhelming national economic interests, or the clearly superior need demonstrated by the very merits of a proposal.

The administrator of the act is always conscious of the possibility of powerful forces effecting a change in the act that may have shattering future consequences. One is always aware of the need to strike a balance, to proceed with caution, to accommodate today in order to prevail tomorrow. Yet the act has been honed and polished in such a way that it need not be compromised. Its strength will grow because the credibility of what it stands for will be enhanced. As life becomes more complicated and the burgeoning needs of the human species impinge with increasing pressure on nonhuman species, the need for informed decisions becomes more acute. Good science must be accompanied by a better public understanding of how our lives touch other species—and how their well-being affects our own.

Of course, we find our lives more complicated by this understanding. It is a blade with two edges. On one side there is a temptation to administer the law through political compromise, more often than not unwarranted. For example, a species considered to have little intrinsic merit may affect a major project (such as a dam or highway) to the extent that it cannot proceed without major adjustments or serious consideration of some complex alternative. This can excite the interest of politicians who are sensitive to the idea that human constituents are far more important than any other species, especially when a species does not enjoy a great deal of popularity. Protection of certain fishes of the Colorado River, for example, has increasingly come into conflict with the need for increasing exploitation of that river. In order for the fish to survive, water must be made available at the right time and in the right quantity. Moreover, temperature, rate of flow, and other physical characteristics of the river are often crucial habitat components. It is difficult to know what to do: Requisite studies are expensive and take many years to conduct, and water development projects— or at least their proponents—cannot wait. In such a circumstance, politicians often bring intense pressure to bear upon the administrators of the act, even to the extent of threatening legislative action to "fix" the problem. Administrators, ever conscious of the power of politicians, may try to resolve the issue by offering mitigation schemes designed to obviate the hazard to a species with the understanding that any adverse consequences to the species will be taken care of at a later time.

Yet to succumb to the temptation to fix the problem after the fact is to admit an inability to deal with the issue now, when there are many possible options. As a project moves forward, the number of choices available for mitigating the impact on a threatened or endangered species becomes limited or even nonexistent. The idea of finding a convenient way out today, the cost of which may be extirpation or extinction, is a denial of the purpose and intent of the act and establishes a dangerous precedent.

On the other side, some may be drawn to use the ESA's

strength to stop events that they oppose. It is appealing to "find" a species and use it to forestall an action of which one does not approve and then, once successful, to forget the species. Both of these stratagems are inappropriate applications of the law. Each undercuts the true value, even the sanctity, of the other species with which we are inextricably allied.

In retrospect, viewed by one who has administered the act and now looks at it from the perspective of a citizens' conservation organization, the Endangered Species Act has worked at two levels. On one level the act is a statute designed to create an institutional sensitivity toward all species and to provide the greatest possible protection for those found to be at hazard of extinction. On another level, the act has allowed us to gain a better understanding of our species' relationship to all the others with whom our destiny is entwined. We have been obliged to face the reality that what we human beings do has a bearing on the well-being of other species, and, most important, that our own well-being is in large measure determined by how other species fare. We are, perhaps, on the verge of recognizing the absolute necessity to make decisions so there are no losers, even if it means our species must accept compromises of a kind we seldom thought about fifteen years ago.

Every retrospective exercise implies a prospect. The noble aspirations of the Endangered Species Act as it was formed over fifteen years ago will prevail only if there is a commitment to vigilance and to assuring there is no steady erosion of the act because we are too impatient or inordinately greedy. There must be resistance to letting the act be used to solve problems more properly resolved by other means, to urgings that the act is "inconvenient" for some, or to the attacks by those who continue to find it an impediment to business as usual.

Once in a while a collective decision is made that things must be done right—that we have an obligation transcending the usual day-to-day living of life. That obligation may be to the future, to an oppressed or disadvantaged few, or to an ideal. The Endangered Species Act represents a national pledge to the future—to an emerging understanding of the role we humans play in the transactions of the natural world. If we remain alert, caring, and committed, the prospect for a body of enlightened legislation, as well as for the reality it addresses, is bright.

LOOKING BACK OVER THE FIRST FIFTEEN YEARS

by

MICHAEL J. BEAN

ON APRIL 29, 1988, a California condor hatched in the San Diego Wild Animal Park. This event, the first such hatching ever to result from condors mated in captivity, was widely hailed as a dramatic breakthrough in the decades-long struggle to prevent the extinction of this largest of North American birds. That success followed by only a few months the birth in Wyoming of the first litter of black-footed ferrets to survive more than a few days after being born in captivity. Amidst the celebration of these conservation milestones, it may be forgiven if a few people temporarily forgot that years of virtually unparalleled effort and expense to preserve wild populations of these two species had already ended in failure. Captive propagation, once intended as a tool to supplement other conservation efforts, now represents the only hope for these species.

The examples of the condor and ferret underscore the difficulty in assessing the results of the Endangered Species Act. Are they success stories or are they failures? Unsatisfying as the answer may be, they are not yet either. Rather, they are still unfinished stories. That they are packed with suspense and tragedy is already apparent; whether their endings will be happy or sad cannot be foretold. The same can well be said of the Endangered Species Act itself.

When the act was passed by Congress and signed into law by

37

President Nixon, conservationists heralded it as a turning point in our relationship with the other living creatures with whom we share the earth. Motivated by the sobering recognition that "economic growth and development untempered by adequate concern for conservation" had driven numerous species to extinction and endangered many others, Congress boldly sought to stem the tide of extinctions depleting the diversity of life itself.

Today, more than fifteen years later, how much of the act's promise has been realized? On the encouraging side, nearly every state has enacted its own endangered species legislation and established its own program paralleling and supplementing the federal program. Concern for endangered species has been integrated into the programs of most federal agencies, including the vitally important federal agencies that manage one-third of the nation's land. The Endangered Species Act has also stimulated major conservation initiatives by private organizations like The Nature Conservancy, which has used its unique land acquisition talents to acquire and protect habitats for many endangered species.

The ultimate measure of success or failure of these efforts, however, is whether the species that are the objects of the act's concern face a more or less secure future. For some, prospects for survival are definitely brighter than they were fifteen years ago. The bald eagle, symbol of the nation, is making an encouraging comeback all across the country; the peregrine falcon, once completely extirpated from the eastern United States, has been successfully reintroduced there; the brown pelican in the Southeast and the American alligator throughout the South have fully recovered; even the whooping crane, reduced to only fifteen birds in 1941, has now been increased more than tenfold.

If the number of recovered species seems few, it must be remembered that fifteen years is a very short time in which to expect dramatic results. During that period, however, the foundations for future recoveries have been laid. For many species, the likelihood of eventual recovery has increased because research done under the Endangered Species Act has made it possible to understand better the causes that threaten their survival and to identify the actions needed to remedy these threats. For others, we may only have bought additional time. Additional time is no small matter, however, for it may prove to be vital time in which to design more long-lasting solutions.

But the negative side of the ledger is not small. Efforts to protect the last remaining wild California condors and black-footed ferrets have failed; their future hinges entirely upon the success of captive breeding efforts. The Palos Verdes blue butterfly of California no longer has even that hope. It went extinct earlier this decade despite years of formal protection under the Endangered Species Act. So too did Florida's dusky seaside sparrow, a species first listed for protection in 1967 under the original Endangered Species Preservation Act of 1966. The last individual of that species died in captivity on June 16, 1987. Outside our nation's borders, the situation is even more bleak. The African elephant, rhinoceros, giant panda, and chimpanzee, species to which major conservation efforts have been directed, continue to spiral downward. As yet undescribed species disappear daily from the relentless pace of deforestation in the tropics.

If any lesson is clear after seventeen years of experience under the Endangered Species Act, it is that the threat of extinction is far greater than it was appreciated to be in 1973 and that the resources needed to address the problem are far greater than those that have been made available thus far. Significant sums have been spent to aid the conservation of the condor, ferret, bald eagle, and a few other species. But these represent a tiny fraction of all the species now formally protected under the Endangered Species Act. In early April 1988, just a few weeks before the baby condor's birth, that list surpassed the one thousand mark. At least that number have already been identified for possible future listing.

Are the resources needed for effective conservation of these species likely to be made available? Fortunately, public support for endangered species conservation, at least as expressed in public opinion polls, is high. Recognizing the strength of that support, elected officials are always at pains to emphasize how much they favor the protection of endangered species. Unfortunately, these same officials often make their loudest declarations of general support just as they are about to propose pulling the carpet from beneath efforts to protect a particular species. In congressional debates, proponents of amendments to strip the government of authority to reduce the drowning of endangered sea turtles in shrimp nets, to lessen the rigors of the act's restrictions with respect to federal highway building, to remove a

species altogether from any protection under the act, or to otherwise frustrate protection efforts can invariably be counted on to
begin their remarks with some version of "I am a strong sup-
porter of protecting endangered species" (usually accompanied
by a grandiloquent reference to the majesty of the bald eagle).
Most of them follow that introduction with an all-important
"however" (typically accompanied by a sneering reference to
the snail darter).

To understand this seeming contradiction, one must consider
the political history of the Endangered Species Act. That history
can usefully be divided into two eras. The watershed dividing
them was the great tumult precipitated by the battle over the
snail darter and the Tellico Dam project. Until that event, little
controversy attended the federal endangered species program.
The procedures for adding species to the threatened and endangered lists were fairly simple, and proposals to list particular
species seldom met with strenuous objection. The act's all-
important Section 7, requiring federal agencies to ensure that
their actions do not jeopardize the continued existence of any
species, had not precipitated any major controversies. Protecting species threatened by extinction was perceived, both by the
public and by its elected officials, as a good thing—or at least a
harmless thing.

Then came the test. The federal courts (ultimately including
the Supreme Court), the Congress, newspaper editorial writers
throughout the country, and countless others all wrestled with
the same question: Which was more important—to prevent the
extinction of a fish that virtually no one had ever heard of or to
build another TVA dam that virtually no one had ever heard of
either? Editorial writers either ridiculed the notion that protecting a mere fish could justify scrapping a multimillion-dollar
dam or rushed to point out that the dam was a wasteful expendi-
ture of tax dollars regardless of its environmental impact. Im-
plicit in the latter argument, of course, was the not very
reassuring notion that a truly worthwhile dam would clearly be
too important to sacrifice for a mere fish. The Supreme Court did
not have to wrestle with these value judgments; its role was
simply to discern the will of Congress. Congress clearly meant to
save the fish, not the dam, said the court. Congress, however,
had the last word and said otherwise.

Since that upheaval, life has not been the same—and not just for the snail darter. Congress and the development community learned very quickly that the Endangered Species Act could lead to major practical consequences and that these might be unpopular and costly. Suddenly, elected officials began to add a "however" to their declarations of support for protecting endangered species. They also added a host of complicating amendments to the Endangered Species Act, focusing in particular on the listing process in order to slow down the addition of new species for the act's protection and thereby reduce the number of potential conflicts in the future. Whereas new listing proposals once generated little controversy, now a proposal is truly unusual if it fails to generate controversy. Once a species is listed, the federal agency responsible for its protection is rarely able to escape political battering, cajoling, threatening, and worse—all aimed at keeping the agency from being too vigilant in carrying out its duties for that species.

The political pressures have often been too much for the Fish and Wildlife Service to bear. One cannot escape noticing the irony in an FWS report that a recent survey turned up none of the four endemic Tombigbee River freshwater mussels the service listed as endangered in 1987. Completion of the Tennessee-Tombigbee Waterway effectively sealed their fate; only after that project's completion did the FWS conclude that it was safe to list these species, whose obituaries can now be readied for future publication. Listing of the Alabama flattened musk turtle was also delayed beyond the deadlines specified in the act, while the FWS reportedly assured the state's congressional delegation that the listing of the turtle would never be the basis for clamping down on water pollution from the coal industry. A regional FWS determination that construction of Stacey Dam in Texas would jeopardize the survival of the Concho water snake was quickly reversed by Washington after congressional pressure. Not needing to await a headquarters directive, the Denver regional director reportedly put out the word that he would insist on a no-jeopardy ruling for Denver's controversial Two-Forks Dam project even before the service's biological studies were completed. And in Arizona, two Fish and Wildlife Service biologists have testified that they were instructed to find that the Mount Graham red squirrel would not be jeopardized by construction of an observatory in its only habitat. These examples, and many others like them, reveal the disquieting side of the statistics so often cited to show that, since Tellico Dam, there have been virtually no conflicts between endangered species needs and development desires.

Political pressures and the necessity to accommodate at least some of them may well be inevitable in any program with regulatory consequences. The danger inherent in such accommodations, however, is that if they are made too easily and too often, they create a perception that the agency charged with administering the program is willing to abandon its basic duties to escape political heat. This perception, in turn, emboldens still others to pressure the agency for even more concessions. It is this dilemma that now confronts the Fish and Wildlife Service and the federal endangered species program. It is, at bottom, the reason why a handful of senators for three years blocked Senate consideration of legislation to reauthorize the Endangered Species Act over issues of no immediate importance to the vast majority of the Senate.

The challenge for the endangered species program in its next fifteen years, and particularly during the current administration, will be to restore the perception that decisions in the program are in fact being made on the basis of the scientific criteria that the law demands rather than in response to political pressures. To restore that perception, the Fish and Wildlife Service and the National Oceanic and Atmospheric Administration must be led by dedicated figures who are broadly experienced in the management and conservation of living resources, strong in their conviction that it is the job of these agencies to base their decisions on the best scientific data available, determined to seek the increased budgetary resources needed for an effective endangered species program, and widely regarded as having unquestioned integrity.

Today, the first member of a new generation of California condors sees a world of bright lights, human faces, and cage bars. Whether it and others that may follow will ever again see a world of rugged mountains, distant horizons, and open skies depends upon how committed the stewards of the endangered species program are to achieving those ends. For the condor, that commitment appears to exist. For the success of the endangered species program, no less a commitment must be made for the many other plants and animals whose very survival is at stake.

LIFE IN JEOPARDY ON
PRIVATE PROPERTY

by

HOLMES ROLSTON III

THE ENDANGERED SPECIES ACT of 1973 is one of the most excit-
ing measures ever to be passed by the U.S. Congress, perhaps to
be passed by any nation. On the surface, the act might seem
analogous to other environmental legislation—about clean air,
soil, and water or timber and range management—in that it
protects people from harm done by other people. But the Endangered Species Act, which indeed protects people from harm,
mandates protecting nonhuman species, whooping cranes and
whorled pogonias. The U.S. Supreme Court, interpreting the
act, insisted that "Congress intended endangered species to be
afforded the highest of priorities, . . . and that the plain intent of
Congress in enacting this statute was to halt and reverse the
trend toward species extinction, whatever the cost."[1]

Implementing this legislation has proved to deepen our range
of moral concern—more so, perhaps, than Congress and the
original supporters of the act expected. We have to ask, "What is
it about species that is of value?" And that takes us deeper into a
concept of natural value than does most natural resource legislation, past the usual economic and utilitarian values attached
to natural resources. The act values rare species in nontraditional and hitherto unstated ways. It challenges property rights
with a higher priority afforded to endangered species. But more
than that, past even a deep humanistic concern, we have to ask

43

how natural things are of value, not just what other humans value. We have to ask whether animals, plants, species, and ecosystems count morally.

The Endangered Species Act laments the lack of "adequate concern" for vanishing species. But neither politicians, nor scientists, nor ethicists have fully realized how developing this concern requires an unprecedented mix of politics, biology, and ethics. To some extent, an act forged in a pluralist democracy needs only to recognize a consensus that we ought to save endangered species; it need not detail the reasons why, on which we may not be agreed. Nevertheless, injunctions coupled with reasons are more likely to succeed, especially if the injunctions are prohibitive. As the act takes on history, Congress, agencies, courts, and citizens must think through the logic of an adequate concern. In so doing, three areas are of primary concern: natural values versus economic values, evolving concepts of private property as they relate to endangered species, and distinguishing moral concern for species from that for individuals.

NATURAL VALUES VS. NONECONOMIC VALUES

The Congress finds and declares that:

1. Various species of fish, wildlife, and plants in the United States have been rendered extinct as a consequence of economic growth and development untempered by adequate concern and conservation.
2. Other species of fish, wildlife, and plants have been so depleted in numbers that they are in danger of, or threatened with, extinction.
3. These species of fish, wildlife, and plants are of esthetic, ecological, educational, historical, recreational, and scientific value to the Nation and its people.[2]

All three clauses contain value judgments mixed with facts. (1) Extinction is a fact; inadequate concern is a value judgment. (2) Danger of extinction is both a fact and a negative value. (3) Many citizens value these species; that fact reports what such persons value. Do the values defended in clause (3) supplement or constitute the values in (1) and (2)? Clause (3) is integral to this act of "the nation and its people," but suppose that it were missing. Clauses (1) and (2) would still lament inadequate concern and the resulting threat and danger to species. Humans are censured for inadequate concern; species are the losers. That in itself might be bad *without* clause (3) about the additional loss to humans. Add the loss to humans, and an act of Congress is in order. Or is it only the human loss that drives all the concern? Were it not for this loss, Congress would not care about species. The act can be read either way. Further, some of the values in clause (3)—for example, the ecological and scientific ones—may include humanistic and naturalistic components inseparably entwined.

Note that "economic" does not appear in the list of values to be protected; to the contrary, it appears counter to the list. Congress wants to "temper" economic growth and development in order to prevent danger, threat, and extinction and to protect aesthetic, ecological, educational, historical, recreational, and scientific values. In this concern for noneconomic and even nonhuman values, we have an extraordinary natural resource law.

In the 1978 amendments to the Endangered Species Act, Congress provided for a multiagency committee to balance economic interests with noneconomic values, but it used great caution lest economic interests prevail easily.[3] That this committee was nicknamed the "God Committee" indicates the high order of proof required for exemption. In the 1982 amendments (reaffirmed in 1988 against a motion to repeal), Congress insisted that the decision to list or delist a species must be based on biological evidence rather than economic effect.[4]

Yet despite the careful language of the act, the argument most often given for conserving endangered species is that some of them (which ones we may not now know) will have economic uses in the future. In the Convention on International Trade in Endangered Species of Wild Fauna and Flora (CITES), Congress constrains trade to protect the increasing cultural and economic values of endangered plants and animals. The International Union for the Conservation of Nature and Natural Resources says: "The ultimate protection of nature . . . and all its endangered forms of life demands . . . an enlightened exploitation of its wild resources" (Fisher and others, 1969, p. 19). Norman Myers concludes that "if species can prove their worth through their contributions to agriculture, technology, and other down-to-earth activities, they can stake a strong claim to survival space in a crowded world" (Myers, 1979a, p. 56). Elsewhere he urges "conserving our global stock" (Myers, 1979b).

That is not what Congress says at all. It does not say: Save those species that are economically valuable. To the contrary, Congress says: Temper economic growth by saving species that have other kinds of values. The title is not "An Act Conserving Our Global Stock" or "An Act for the Enlightened Exploitation of Species." The Endangered Species Act is a congressional resolution that the nation and its people ought to live as compatibly as they can with the fauna and flora on their continent (and abroad), and it deplores the fact that we are not now doing so. The act claims that what is good for the fauna and flora is good for the people. Anyone who thinks the contrary has the burden of proof.

Working out this general principle will mean, at the level of particulars, that somebody, somewhere, will have to "temper" their economic interests lest they endanger species and lest people lose these noneconomic values that the act intends to protect. But what does it mean to temper economic growth out of concern for these deeper natural values?

EVOLVING CONCEPTS OF PROPERTY

Fauna, Flora, and Landowners

The Endangered Species Act, people first think, is about grizzly bears and bald eagles. Animals and birds move around. They live in dens and nests in particular places but range over hundreds or thousands of square miles. In the case of big animals and migrating birds, it is easy to see that they do not belong to a local landowner or even a state. There is a long legal tradition that property holders do not own vertebrate wildlife, even if such wildlife resides entirely on one owner's property. Wildlife is a common good held in trust by the state for the benefit of the people.[5] Hence, although the landowner does control access to his land, the state decides when animals can be taken and who is licensed to do so. This legal tradition arose with regard to individual animals, but the welfare of the species has figured into such regulations for both game and nongame species. No great stretch of thought is required to agree that all the animals, birds, fish, perhaps even the butterflies and the bees, are common goods that can be regulated by the state.[6] The prohibition of the Endangered Species Act against taking animals on private land, including invertebrates, has not been seriously challenged. Further, the federal government can designate critical animal habitat on private land.

Yet the Endangered Species Act, people soon discover, is not only about fauna but also about flora—pitcher plants and cacti. In the original 1973 act, plants were already of concern; one could not engage in unrestricted interstate commerce in protected plants. Plants could be listed regardless of where they occurred. Initially, however, there was no prohibition against taking endangered plants on public land, much less private land. Since 1973, Congress has increasingly expanded protection for endangered plants. In 1982, Congress prohibited the private collecting of endangered plants on federal lands; technically this did not prohibit destroying them by logging, grazing, building dams, and so forth, as long as one was not collecting them.[7] The 1988 reauthorization of the act makes it a violation maliciously and knowingly to damage endangered plants on federal land. That ban prevents vandalism, but it may not require people to notice what their cows eat or whether their ORVs crush protected plants. The 1988 amendments also prohibit harming endangered plants on private land in violation of state law, including trespass law.[8] State laws vary widely; some states do indeed prohibit anyone, landowner or not, from taking endangered plants on private property without state permit. Most states, however, require nonlandowners to obtain a permit to collect or destroy an endangered plant, but allow landowners to do what they please.

Plants do not move around, though some disperse their seeds on the wind or in the fur of animals. Rare plants move less than common ones. Traditionally, the trees and grass belong to the landowner. "Standing timber is real estate. It is a part of the realty the same as the soil from which it grows" (*Kerschensteiner v. Northern Michigan Land Co.*).[9] Except for noxious weeds and marijuana, the state has made few claims regulating flora on private land. Plants on public land can certainly be protected by federal and state law, since the government owns (or is responsible to the people for) those plants—but what about plants on private land?

The pine trees in a southern swamp belong to the landowner, but how about the plants of *Rhododendron chapmanii*, a listed species? Individual plants, we have always thought, are part of the property; but we face a new question when a few individual plants constitute a natural kind. Ownership of a species has never been part of the explicit bundle of property rights. Who owns a species? This question has remained tacit with wildlife because landowners do not own the individual animals, much less the species. But with plants the question comes into sharper focus.

Taking Property and Taking Species

In our efforts to implement the Endangered Species Act, we are confronted with a double use of "take"—where the "taking" of life and species is set against the "taking" of property. In struggling toward resolution, our moral and legal convictions about the institution of private property and its economic value are evolving in the encounter with the biology and ecology of endangered species. First, we should try to appreciate how things look from the perspective of the landowner. Some landowners are excited to have endangered species on their property, as long as it does not conflict with their use of the land. Other landowners, finding a conflict between their preferred uses and preservation, think it a misfortune. One of three known populations of the San Diego mesa mint, a tiny endangered plant, was deliberately destroyed by a private developer to ensure that subsequent requests for federal construction grants would not be delayed (Carlton, 1986). This is not an isolated occurrence. Bruce Mac-Bryde (1980, p. 33), a botanist for the U.S. Fish and Wildlife Service, found that several candidate endangered plants on private lands have been intentionally destroyed in the last few years.

The Endangered Species Act sets untempered economic growth against adequate care and concern for aesthetic, ecological, educational, historical, recreational, and scientific values. But where rare species occur on private property, it sets concentrated economic benefits to the single landowner against diffused general benefits to citizens. The landowner, also a citizen, shares in these benefits, but gains only a soft set of benefits against heavy costs in opportunities forgone. The nation and its people enjoy the claimed benefits without cost, but the landowner, constrained in the right to hold and enjoy property, suffers economic loss. The benefits desired by the landowner are appreciable, immediate, apparent, quantifiable, typically economic. The benefits to human beings as a whole are dispersed and delayed, typically noneconomic, though in the aggregate they might outweigh benefits to the landowner. Moreover, the landowner does own the land, which makes the case different from that of entrepreneurs who wish to turn a profit on public lands.

Every state regulates the ways in which property owners can develop their land. The more special and sensitive the land (a floodplain, a coastal zone, a wetland, open space, a scenic or historically significant place), the tighter the regulations. Few persons own real estate without zoning and other ordinances that restrict the ways in which they can use it. These restrictions protect public goods. As such, there is no contesting the state's power to regulate. One can, however, contest the extent of regulations. At some point, land-use restrictions can amount to confiscating private land for public use—a benefit for which the public ought to pay.

The question that arises is this: Do prohibitions against destroying plants on private land also involve a taking of property that requires just compensation under the Fifth Amendment to the U.S. Constitution? The government is prohibited from "taking" private property without showing just cause (that is, a benefit to the public which outweighs the disadvantage to the unwilling landowner) and without fair compensation. Thus the distribution of benefits and costs will involve an unwilling landowner, but at least fair compensation will distribute benefits and costs equitably.

The word "take" also occurs in a newer, significant context. In the Endangered Species Act and its amendments, there are frequent prohibitions against "taking" species that are listed as threatened or endangered. The word "taking" comes from a legal tradition where one "takes possession" of property and where, in a parallel use, wildlife is free and belongs to no one until a (licensed) hunter "takes" or captures it whereupon the animal becomes his possession. In the original act, taking restrictions applied only to animals, but the prohibition has been increasingly applied to plants (McMahan, 1980, pp. 562, 564). Mixed with ownership, the word becomes a euphemism for taking life—for killing (or at least capturing or uprooting) an animal or plant and simultaneously taking the species it instances.

In the decade and a half since Congress first passed the Endangered Species Act, the meaning and scope of "take" have been explored, fought over, and enlarged. In the 1973 act, "the term 'take' means to harass, harm, pursue, hunt, shoot, wound, kill, trap, capture, or collect, or to attempt to engage in any such conduct."[10] The definition has animals in focus. Elaborating, the act permits certain kinds of taking provided that they do not jeopardize the species. Additionally, it notes that taking habitat may be tantamount to taking species.

In the 1988 amendments regarding the protection of plants, it is unlawful to "remove and reduce to possession" or to "maliciously damage or destroy" listed plants on lands under federal jurisdiction and "to remove, cut, dig up, or damage and destroy any such species on any other area in knowing violation of any law or regulation of any State."[11] In a suggested model plant act for states, "take" means "pick, collect, cut, transplant, uproot, dig, remove, damage, destroy, trample, kill, or otherwise disturb," but this language has only partially found its way into state legislation (Fitzgerald, 1986).

Through Section 7 of the ESA, the federal government prohibits its agencies from "taking" animal or plant species on public as well as private lands. In theory, the government does this not only at home but also abroad. But can the federal government prohibit a private citizen from taking plants on his or her own land or, with a landowner's consent, on the private lands of others?[12] Congress pragmatically has deferred the question to state law since property rights and regulations traditionally are thought to be more appropriate at the state level. At the same time, Congress has endorsed state laws that do prohibit taking endangered plants with federal penalties for violators. Increasingly, it seems Congress is concerned about taking plants on private as well as public lands. Moreover, in the 1988 amendments to the act Congress directed the Fish and Wildlife Service to develop recovery plans without regard to taxonomic classification, indicating a concern for plants equal to that for animals.

Loss or Gain?

If the public is gaining a good, government ought to compensate; but if government is protecting from harm, it can prohibit without compensation.[13] People have a duty not to harm regardless. This duty can be legally enforced. Courts can stop a landowner from "engaging in conduct which he ought, as a well socialized adult, to have recognized as unduly harmful to others" (Ackerman, 1977, p. 102). This agrees with a general moral rule that injunctions against malevolence are binding in a strong sense, whereas injunctions to benevolence are weaker. Positive rights to be helped are often optional in a way that negative rights not to be harmed are not. I must not injure a stranger who begs on the streets, but no law requires that I benefit him with a handout. When the law enjoins landowners to protect listed plants on their property, is government protecting against harm or is the public gaining a good? Answering such a question requires us to mix economic and noneconomic values, humanistic and natural values, and politics, biology, and ethics in ways we are only now beginning to explore.

In the title of the act itself, the term "endangered" overshadows everything to follow. The first two opening clauses lament the irretrievable extinction of many species and threatened loss of many more. Section 7, with its "no jeopardy" clause, instructs all federal departments and agencies to take whatever action is necessary "to insure that actions authorized, funded, or carried out by them do not jeopardize the continued existence of such endangered species or threatened species or result in the destruction or modification of habitats of such species which is determined . . . to be critical." All this language suggests a perspective of harming, and we regularly use police power and "protective regulations" against jeopardy, threat, danger, and destruction.

Well, it will be replied, although the terms are maleficent, humans are not endangered; only the plants and animals are in jeopardy. In a noteworthy suit, the palila, a Hawaiian finch, is listed as though it is one of the plaintiffs.[14] Certainly the palila stands to be harmed. On a naturalistic reading of the act, the loss of species is a bad thing in itself. But on a humanistic reading, Congress is concerned about extinction only insofar as it concerns humans; only the humans who were plaintiffs along with the palila really have standing to sue. For humans, aesthetic, ecological, educational, historical, recreational, and scientific benefits are claimed in the act. All these benefits are something gained.

But the better perspective is not gain but loss. Even though the loss of life is suffered by animals and plants, there is danger of loss and threat of serious harm to humans who lose too when these animals and plants vanish. So the human benefits, though not a matter of life and death, are a matter of danger and threat of loss. The U.S. Supreme Court found in the act "repeated expressions of congressional concern over what it saw as the enormous danger presented by the eradication of any endangered species."[15]

Landowners will argue that the things they want to do (cut the timber, build a summer home, fill a swamp) have not hitherto been thought of as harming the public. To the contrary, they have been judged positively. The right to develop one's land has always been a standard property right; most property is in fact purchased for its development potential. But even if the landowner intends to do something worthwhile, the threat of extinction or actual extinction may result. Indirect and incremental harm is still harm. In the state of Hawaii, untempered development threatens more than 700 of the 2,000 to 2,500 plants endemic to the islands—at least one taxon in three or by some estimates half the flora. California, Florida, Oregon, Texas, Utah, Arizona, and Puerto Rico stand to lose plant species in the hundreds, and most states stand to lose more than a dozen. Perhaps 3,000 species, subspecies, and botanical varieties are at risk out of 22,000 known in the United States—about one taxon in seven (McMahan and Walter, 1987; Altevogt and MacBryde, 1977).

Landowners may counter that whether there is gain or loss

depends on who has recently shifted perceptions of values. The goods carried by rare species are novel, nontraditional, public goods; the Endangered Species Act asserts new benefits hitherto unrecognized and unclaimed. But changing perceptions of value may be realizing values that were long in place and accepted as natural givens. Environmental values are not simply in the eye of the beholder. When we lose air, water, soil quality, natural resources, when we lose ecosystem stability, we lose whether we are aware of these losses or not. We lose even if we think we do not. There was loss when the passenger pigeon went extinct, even though this loss at first was perceived by few people. Untempered economic development seriously harms the public because it irreversibly harms life processes. These biological processes are not newly adopted values; they are the oldest values of all. It is no part of a landowner's rights to "take" this life; nor is the state "taking" something from the landowner when it insists that a species be preserved.

Property rights were instituted to protect individuals from harm; now we must institute a law to protect individuals from harming species and in so doing harming other persons. John Locke asserts that the landowner's property rights give him no right to spoil or destroy—"the exceeding of the bounds of his just property not lying in the largeness of his possession, but in the perishing of anything uselessly in it." We can make use of the commons, but we can take only "where there is enough and as good left in common for others" (Locke, 1690 [1947], chap. 5, secs. 46 and 27). In the case of nonrenewable resources, use may leave less for others, but land use ought to be renewable. Land can be left to others. Locke did not have species in mind, but the principle applies here. Species ought to be renewable resources; they are wealth on the land. When species perish, uselessly or not, this creates scarcity. Alternately put, property rights to land ought to help us divide up the pie of natural resources, but extinction of species shrinks that pie forever.

In the case of long-continuing, nonreplaceable goods, property rights are rights to use, not to destroy. I cannot, on my private forested land, cut timber in such a way that, with the soil eroded and seeding stock gone, the forest cannot be regenerated.[16] I can buy real estate and build a home there. Perhaps I must destroy the native vegetation to build my home, but that does not destroy the possibility of replacing the former vegetation with like or other vegetation, nor do I destroy the possibility of other uses of the land subsequently. But I cannot poison the land so that no vegetation can be grown there ever again, nor can I pollute it with plutonium so that no one else can use it for ten thousand years. I can buy a swampland, but can I destroy the rare swamp ecosystem? Or endemic species there? Property rights on land include the right to destroy tokens, but not types. We are beginning to see how protecting nature can be more important and more moral than protecting property.

CONCERN FOR INDIVIDUALS VS. SPECIES

We need the perspective of natural history. True, Congress does not often look after ecological, historical, and scientific values in nature. Nature does not run by act of Congress. But Congress in the 1973 Endangered Species Act laments the lack of adequate concern for endangered species; it worries about irretrievable loss, for not even an act of Congress can remake a species. An act of Congress, however, might save a species; Congress can resolve to let natural history continue. Making such law reasonable may involve our reeducation about what a species is and about what humans are doing to other species. It will involve distinguishing between benefits to individuals—typically sentient and usually persons (the traditional focus of Western ethics and law)—and respect for life at the species level. With plant species, this process may take a generation—but in the last fifteen years we have begun to see this reeducation.

G. G. Simpson concludes, "An evolutionary species is a lineage (an ancestral descendant sequence of populations) evolving separately from others and with its own unitary evolutionary role and tendencies" (1961, p. 153; endorsed for plants in Grant, 1981, p. 83). Niles Eldredge and Joel Cracraft (1980, p. 92) insist that species are "discrete entities in time as well as space." What the nation and its landowners want to respect are dynamic lifeforms—biological vitality that persists genetically over thousands, even millions, of years, overleaping short-lived individuals. Although species are always exemplified in individuals, the species is a bigger event than the individual. The species level is more appropriate for moral and legal concern since the species is a more comprehensive survival unit than the organism. What survives for a few months, years, or decades (rarely centuries) is the individual plant; what survives for millennia is the kind.

When a rhododendron dies, another one replaces it. But when *Rhododendron chapmanii* goes extinct, the species terminates forever. Death of a type is radically different from death of a token. Extinction shuts down the generative processes, a kind of superkilling. This kills forms (species) beyond individuals. This kills essences beyond existences—the soul as well as the body. This kills collectively. To kill a particular plant is to stop a life of perhaps a few years while other lives of such kind continue unabated and the possibilities for the future are unaffected. To superkill a species is to shut down a story of millennia and leave no future possibilities. One generation stops future generations. In this sense, "harm" takes on a profound significance.

What is wrong with human-caused extinction is not just the loss of human resources, but the loss of biological sources. Certainly we care about values to the nation and its people, but we should also care about biological processes that take place independently of human preferences. In former times, humans had neither the power to cause mass extinctions nor the knowledge of what they were inadvertently doing. But today humans (certainly those who support, authorize, and implement the Endangered Species Act) have a greater understanding of the speciating processes, more predictive power to foresee the intended and unintended results of their actions, and more power to reverse undesirable consequences. Increasingly, we know floristic locales and natural histories; we find that willy-nilly we have a vital say in whether or not these evolutionary stories continue. We have sufficient knowledge and control over the threatening social, economic, and political forces to have options. Never before have questions at this level been faced. The answers are generating a deeper sense of responsibility. Humans ought not to superkill a species without a superjustification. This may involve redefining what property rights mean in the light of learning what a species is and discovering the values carried by species.

Some claim that we should protect rare plants that were once common since such plants are likely to serve important ecosystem functions, whereas naturally rare species are less significant. With their loss, the argument runs, less is lost. But naturally rare species, as much as common ones, signify exuberance in nature—each presents a unique expression of prolific evolutionary potential. The rare flower is a botanical achievement, a bit of brilliance, a problem resolved, a niche filled. The local endemic species, perhaps one specialized for an unusual habitat, represents a rare discovery in nature before it provides rare values to the nation and its people. Although it might be true that naturally rare plant species are of lesser ecological value on a regional scale, rarity need not reduce scientific, aesthetic, educational, historical, and recreational values.

If one insists on a restricted evolutionary theory, rare plants are random accidents resulting from mutations and contingencies of natural history. Such a biological origin might seem to suggest little of value, little duty, and little to be lost. But a more comprehensive theory also suggests that surviving plants must be satisfactory fits in their environments. Sometimes they live on the cutting edge of exploratory probing; sometimes they are relics of the past; often neither new nor old, they occupy niches that have always been rare. Whatever the explanation of their rarity, they offer evidence, promise, and memory of an inventive natural history. Even more poignantly than the common, they are a sign of life persisting in struggling beauty, flourishing, pushing on at the edge of perishing. The rare flower—if one opens to a wider, philosophical perspective—offers a moment of perennial truth. Life is a many-splendored thing; extinction of the rare forms dims this luster.

Rhododendron chapmanii, though rare, has proved its right to life, has proved that it is right for life, when tested by natural selection. It is a satisfactory fit in its niche in the transition zone between the dry longleaf pine forests and the moist *Cyrilla* thickets. Some rare plants may be en route to natural extinction, but it does not follow that most have less biological achievement than common species. Endemics or specialized species may competently occupy restricted niches. When the nation and its people encounter such a species, its adaptive fitness in an ecosystem generates a presumption that it is right for us to let them be. A rare plant's roots literally go down in the landowner's land, but its roots historically go back across millennia, long before the land was owned, long before this nation existed. Recent possession conveys no right to destroy species and with it the future. A species is not personal property. The appropriate relation is that of caretaker to a trust rather than property owner to his property.

Indeed, the state is only their recent trustee. The concept of ownership even by the state of wild fauna and flora has gradually eroded and been replaced by a nonpossessive concept of their value and significance, something like trusteeship. This is legally and ethically correct; wild species are not the possession of anyone. This is essentially what "wild" means: outside the possession, control, management, and ownership of humans. In this light, an act of Congress to extirpate a species is outside its legitimate authority. The high-level committee that can permit destruction of species is called the God Committee in a mixture of jest and theological insight. Some things belong to Caesar, and some do not.

In an environmental ethic, there really is no reason to think biologically or philosophically that the flora are more or less part of the land than the fauna. Both are entwined with the landscape. A species is what it is *where* it is. Only legally have we separated the two in the past. While this makes some sense dealing with individual plants and animals, we now must think in terms of species, bringing law into line with biology and ethics.

Respect for life is tacit in the Endangered Species Act but needs to be made more explicit. Several billion years of creative toil and several million species of teeming life have been handed over to the care of this late-coming human species in which mind has flowered and morals have emerged. If we are to be true to our specific epithet, ought not *Homo sapiens* value this host of species as something with a claim to care in its own right? A nation and its people have resolved to form an adequate concern for these species that are endangered by human economic growth and development. For a landowner to assume the right to destroy a species hardly seems biologically informed, much less ethically adequate. Nor will the law of the land be legally adequate until the nation makes it clear that property rights on this land do not override a species that is right for life. Landowners do not own species. Indeed, the only landowners who really possess and enjoy their land, in a deeper philosophical sense, are those who respect the life that is native there. The only people who really possess and enjoy their landscape, their country, are those with an adequate concern for fauna and flora.

Notes

The author thanks Linda McMahan, Faith Campbell, Bruce MacBryde, Kathryn Kohm, and Paul Opler for critical comments.

1. *TVA v. Hill*, 437 U.S. 174, 184.

2. Endangered Species Act of 1973, Section 2(a) (16 U.S.C. §1531(a)).

3. See Endangered Species Act Amendments of 1978; Pub. L. 95-632, 92 Stat. 3751 (Nov. 10, 1978).

4. See Endangered Species Act Amendments of 1982; Pub. L. 97-304, 96 Stat. 1411 (Oct. 13, 1982).

5. This was formerly phrased as state ownership of wildlife, but that concept has been subsumed under the state's power to regulate all natural resources. This has developed as an expanding public trust doctrine, first applied to navigable waters. "The cornerstone of environmental law is the assertion that all of our national natural resource treasures are held in trust for the full benefit, use and enjoyment of all the people of the United States, not only of this generation but of those generations yet unborn, subject only to wise use by the current nominal titleholder. . . . The basic principle underlying the Trust Doctrine is that 'There are things which belong to no one, and the use of which is common to all' "(Yannacone and others, 1972, vol. 1, p. 11, citing *Geer v. Connecticut*, 1896).

6. Sedentary animals (barnacles and clams) are thought to belong to property owners; many of these are marine, however, and ownership in marine waters has been complex.

7. See Endangered Species Act Amendments of 1982; Pub. L. 97-304, 96 Stat. 1411.

8. See Endangered Species Act Amendments of 1988; Pub. L. 100-478 §6, 102 Stat. 2306.

9. 244 Mich. 403, 417, 221 N.W. 322, 327.

10. Endangered Species Act of 1973, Section 3(14) (16 U.S.C. §1532(19)).

11. See Endangered Species Act Amendments of 1988; Pub. L. 100-478 §6, 102 Stat. 2306.

12. The Convention on Nature Protection and Wildlife Preservation in the Western Hemisphere states that the species of fauna and flora listed in its annex "shall be protected as completely as possible and their hunting, killing, capturing, or taking shall be allowed only with the permission of the appropriate government authorities in the country," presumably without regard to whether they are taken on public or private land (56 Stat. 1354, Article VIII, 1366). The United States has not named any plants to the list, though other nations have. The convention has largely been ignored, especially in domestic law.

13. The Supreme Court of Wisconsin found that compensation was not required when a county ordinance prevented landowners from filling a wetland that was a critical natural feature. "We have a restriction on the use of a citizen's property, not to secure a benefit for the public, but to prevent a harm from the change in the natural character of the citizen's property. . . . Destroying the natural character of a swamp or a wetland so as to make that location available for human habitation [degrades] the ecological creation [and] the new use, although of a more economical value to the owner, causes a harm to the general public. . . . The shoreland zoning ordinance preserves nature, the environment, and natural resources as they were created and to which the people have a present right." The court recognized that the public had only recently realized the values in wetlands (*Just v. Marinette County*, 1972, 56 Wis. 2d 7, p. 18, 201 N.W. 2d 761, p. 768). Also, nothing for which the landowners had labored was taken; their loss was of possibilities of development traded against public loss. The ordinance reduced the market value of the land, but that market value did not result from labors of the owners.

    In a New Hampshire decision, the court found that "controlling and restricting the filling of wetlands is clearly within the police power of the State. . . . If the action of the State is a valid exercise of the police power proscribing activities that could harm the public, then there is no taking under the eminent domain clause." The regulation "did not deny to plaintiffs the current uses of their marshland but did prevent a major change in the marshland's essential natural character, a change which plaintiffs, for speculative profit, sought for a purpose unsuited to its natural state and injurious to the rights of others. . . . The plaintiffs' four acres were part of a valuable ecological asset of the seacoast area and . . . the proposed fill would do irreparable damage to an already dangerously diminished and irreplaceable natural asset. . . . The proposed fill would be bad for the marsh and for mankind" (*Sibson v. State of New Hampshire*, 1975, 115 N.H. 124, pp. 124–126 336 A2d 239).

In general, courts have been reluctant to advance a theory of property rights applicable to taking under the Fifth Amendment, preferring a pragmatic approach. A regulation that constrains the owner's desires about the degree and kind of development in order to protect a public good is legitimate, but such regulation ought not to force the owner off the land or to deprive the owner of all reasonable uses of it, especially previously existing uses. If so, it is taking. Many commentators think that the law of taking is inconsistently applied and theoretically confused.

14. *U.S. District Court, Palila v. Hawaii Dept. of Land and Natural Resources*, 1979; 471 F. Supp. 985 (D. Hawaii 1979), aff'd, 639 F. 2nd 495 (9th Cir. 1981).

15. *TVA v. Hill*, 437 U.S. 174, 186.

16. *State of Washington v. Dexter*, 32 Wash. 2nd 551, 202 P. 2d 906 (1949), aff'd per curiam, 338 U.S. 863 (1949).

References

Ackerman, Bruce A. 1977. *Private Property and the Constitution*. New Haven: Yale University Press.

Altevogt, Raymond, and Bruce MacBryde. 1977. "Endangered Plant Species." Pp. 81–91 in *McGraw-Hill Yearbook of Science and Technology*. New York: McGraw-Hill.

Carlton, Robert L. 1986. "Property Rights and Incentives in the Preservation of Species." Pp. 255–267 in Bryan G. Norton (ed.), *The Preservation of Species*. Princeton, N.J.: Princeton University Press.

Eldredge, N., and J. Cracraft. 1980. *Phylogenetic Patterns and the Evolutionary Process*. New York: Columbia University Press.

Fisher, James, Noel Simon, Jack Vincent, and IUCN staff. 1969. *Wildlife in Danger*. New York: Viking Press.

Fitzgerald, Sarah Gates. 1986. "The State of the States in Plant Protection." *Garden* 10(5):2–5, 31–32.

Grant, Verne. 1981. *Plant Speciation. 2nd ed.* New York: Columbia University Press.

Locke, John. 1690 [1947]. *Two Treatises of Civil Government*. New York: Hafner.

MacBryde, Bruce. 1980. "Why Are So Few Endangered Plants Protected?" *American Horticulturist* 59(5):29–33.

McMahan, Linda. 1980. "Legal Protection for Rare Plants." *American University Law Review* 29:515–569.

McMahan, Linda, and Kerry S. Walter. 1987. "Where the Rare Plants Are." *Newsletter of the Center for Plant Conservation.* 2(3):6.

Myers, Norman. 1979a. *The Sinking Ark.* Oxford: Pergamon Press.

Myers, Norman. 1979b. "Conserving Our Global Stock." *Environment* 21(9):25–33.

Simpson, G. G. 1961. *Principles of Animal Taxonomy.* New York: Columbia University Press.

Yannacone, Victor J., Jr., Bernard S. Cohen, and Steven G. Davison. 1972. *Environmental Rights and Remedies.* 2 vols. and supplement, 1988. Rochester: Lawyers Cooperative Publishing Co.

SNAIL DARTERS AND PORK BARRELS REVISITED: REFLECTIONS ON ENDANGERED SPECIES AND LAND USE IN AMERICA

by

GEORGE CAMERON COGGINS

THE TITLE OF THIS ESSAY derives from an article on endangered species and land-use conflicts that a former student and I wrote in 1981. After describing the statute and surveying the cases, we ambiguously concluded:

> The preservation of species facing extinction takes precedence over all other considerations until such time as the [Endangered Species Committee] or Congress declares otherwise.
>
> Given the present and probable future numbers of listed species, the flexibility and uncertainties in the listing and designation processes, the degree of human commitment to either preservation or projects, and the nearly infinite number of instances in which conflicts can occur, matters surely are not as simple as the preceding sentence indicates. The Endangered Species Committee will grant exemptions. The Fish and Wildlife Service and the National Marine Fisheries Service will refuse to list or to issue negative biological opinions. Environmental organizations will neglect to sue. Courts will seek equitable solutions in spite of *TVA v. Hill*. Other perceived needs . . . will take precedence. Nevertheless, endangered species protection will be a significant influence on land

use and related activities in the United States throughout the foreseeable future. [Coggins and Russell, 1982, p. 1525]

That conclusion still holds.

After recounting some opinions about the nature and context of the Endangered Species Act (ESA), this essay surveys litigated instances of its application to disputes arising on federal public lands. We then descend from the global to the local: Two controversies close to my home illustrate how the act has broad and deep consequences for land use in America.

THE ESA AS AN ELEMENT OF WILDLIFE LAW

One of the most fascinating aspects of wildlife law is that it concretely expresses changing public attitudes about the natural world—the universe of ecological processes otherwise unaffected by human intervention. Throughout most of human history, animals (like trees) were thought to exist only to support human life in ways that humans determined.[1] In this country, market hunting and similar forms of wasteful slaughter were the inevitable by-products of this attitude, especially as it was exacerbated by our laissez-faire frontier philosophy. Realizing that greed was killing the golden goose along with bison and passenger pigeons, people eventually demanded an end to the unregulated commons. The systems of controls developed by the early state fish and game agencies—the bureaucratic reincarnations of the Sheriff of Nottingham—were premised on utilitarian ideals. The pioneer wildlife management professionals regarded game animals, fish, and birds as "crops" to be nurtured during breeding seasons and then "harvested" when mature. These primitive attempts to avert the tragedy of the commons manifested societal common sense: If we kill all the deer, there will be none to hunt next year.

Utilitarianism is still the prevailing philosophy in mundane state hunting and fishing regulation. Many people and much law still regard some wildlife species as deserving only of eradication. (See generally Coggins and Evan, 1982.) But the Endangered Species Act of 1973[2] marks both the culmination of a new public mindset and the beginning of a new era in relations between Americans and the supposedly lesser orders.

While the ESA can be justified on practical economic grounds to an extent, simple utilitarian justifications miss the point. The people and their Congress, by instituting legal protection for declining wildlife species, evinced an almost religious reverence for the sanctity of life in all its forms. In 1918, Congress acted to protect birds because they sang prettily;[3] in 1940, Congress singled out the bald eagle for preservation as the living national symbol.[4] By 1973, Congress had dispensed with species-by-species evaluations of good and bad in terms of value to *Homo sapiens*. It simply said that all fauna and flora species of whatever utility were entitled to continued existence.

As law, the ESA is even more radical than the philosophy it embodies. Statutes allocating or protecting natural resources traditionally have been imprecise and flexible.[5] They typically turn over a perceived problem to an administrative entity and instruct it only to go and regulate in the public interest.[6] The ESA, by contrast, is relatively precise and absolute. When its provisions are triggered, the answer usually is clear, and the federal offices and agencies must overcome their propensities to factor in economics, politics, and like considerations. In this respect, endangered species law is unique in public natural resources law.[7]

THE ESA ON THE FEDERAL PUBLIC LANDS

Federal public lands are a critical arena for testing the success of the Endangered Species Act.[8] Even excluding lands used for military and governmental purposes, the United States still owns nearly a third of the nation's entire surface area—more than 700 million acres of mostly undeveloped land. As one might expect, the federal holdings often provide the best (and sometimes the last) habitat for wildlife species in danger of extinction. (See Swanson, 1978, p. 428.)

The federal public lands not used for normal governmental purposes such as post offices and courthouses are in the charge of four separate land management agencies, each of which has a somewhat different attitude toward wildlife preservation. The

Fish and Wildlife Service (FWS) runs the National Wildlife Refuge System and consults with the other land management agencies on compliance with the Endangered Species Act. (See generally Reed and Drabelle, 1987.) The FWS's refuge management is infused with the utilitarian hunting-is-a-valuable-management-tool philosophy (Coggins and Wilkinson, 1987, pp. 828–830), and the FWS has not always been immune to political pressure from developers.[9] Nevertheless, the FWS is the nation's primary wildlife cop and its pro-wildlife influence has spread rapidly. (See generally Reed and Drabelle, 1987.) Of the land management agencies, the National Park Service (NPS) is the most devoted to "naturalness" in managing wildlife. The NPS long ago decided to let the wildlife in parks fend for themselves so that local evolution would take its natural course. (See Coggins and Evan, 1982, pp. 837–840.) Recently, however, the NPS's hands-off policy has come under attack as flawed if not counterproductive (Chase, 1986).

The Forest Service and the Bureau of Land Management (BLM), the "multiple use" land management agencies, are somewhat less solicitous of wildlife welfare (Coggins, 1982). Together they manage nearly a quarter of the nation's surface area. Although the Forest Service is a respected and relatively sophisticated agency, it is dominated by foresters who tend to be concerned more with timber harvest levels than obscure birds.[10] The Forest Service's orientation toward "even-aged management," which encompasses a variety of practices, notably clearcutting, that harm wildlife populations, has brought it into conflict with ESA objectives in several situations. The BLM is the poor relation of the federal land agencies. Its policies historically have been antithetical to wildlife preservation. (See, for example, Coggins and Lindeberg-Johnson, 1982.)

Here we concentrate more on the outcome of litigation than on internal agency procedures and attitudes for the reason that the threat of judicial reversal usually influences an agency's actions far more than good intentions. Lawsuits are devices of limited utility that require vast expenditures of time, effort, and resources, and defenders of endangered species may rely too heavily on them. But it cannot be gainsaid that litigation has shaped endangered species law in ways that were only imagined in 1973. Without judicial review, the ESA could have become just one more paperwork burden without much substantive significance.

The scorecard of wins and losses in lawsuits brought to enjoin potentially harmful projects is not an adequate measure of the ESA's impact on federal land use. Assuming, unfairly, that the federal land management agencies would always prefer economic development to wildlife protection, a more accurate gauge of impact is the changes a proposed project undergoes from initial conceptualization to realization. Two cases decided in 1982 are instructive. In the first, the Forest Service approved a mining exploration plan for an area thought to be inhabited by grizzly bears.[11] In the end, the reviewing court allowed the prospecting operation to proceed, but only because the Forest Service and the FWS imposed a long series of mitigating conditions on the prospector designed to avert any substantial disturbance of the threatened species. In *North American Wild Sheep*,[12] on the other hand, a different court was not convinced that mitigation measures would be effective in avoiding harm to a nonlisted species and enjoined the road reconstruction proposal at issue.

Three lessons stand out from these two cases. One, judges have different opinions just like everyone else. Two, inclusion on the list of endangered and threatened species does not automatically mean that the species' territory is off-limits to all human activity. Three, and most important, agencies realize that the days of economic development *über alles* are over. The project must be shaped to suit the requirements of the listed wildlife or the project, not the wildlife, will be dead in its tracts.

The most important ESA lawsuits are only now being filed and decided as I write in 1989. Since the *Snail Darter* opinion more than a decade ago,[13] most cases have concerned the effect of a single project on a single vulnerable species. The judicial results have been mixed. In cases involving federal lands, courts have remanded agency decisions for failure to observe required ESA procedures in a variety of circumstances, from road building[14] to oil and gas drilling[15] to questionable local mitigation promises.[16] Equally telling, the courts have invariably upheld land management agency actions intended to protect listed species even though the effect on human economic and recreational interests was substantial.[17] In several cases, on the other hand, courts have ruled that the agency sufficiently complied with the ESA even though some harm to the species might result.[18]

That the scope of litigated disputes is widening illustrates the growing impact of the ESA. Perhaps the most significant is the *Minnesota Wolf* decision, in which the Eighth Circuit Court outlawed a sport hunting season on that threatened species.[19] That opinion should have augured well for the grizzly bear—another predator under relentless attack by development in general and human predators in particular. But that has not yet been the case. In the *Sunrise Hunting*,[20] *Steel Shot*,[21] and *Strychnine*[22] cases, federal agencies had to take their general regulatory policies back to the drawing board because they had neglected to ascertain or take adequate account of the effect of the regulations on endangered or threatened species.[23] In the *Nebraska v. REA*[24] and *Riverside Irrigation District*[25] cases, the courts held that the ESA protected the stopover habitat of endangered birds many miles downstream from the projects at issue. In the *Stampede Dam* lawsuit,[26] the court agreed that the Interior Department must protect the spawning grounds of endangered fish from water diversions.

A new wave of litigation promises to affect natural resources management on federal lands in even more drastic ways. The red-cockaded woodpecker and northern spotted owl disputes threaten to turn Forest Service (and BLM) timber production programs inside out. A federal district court has already enjoined clear-cutting and road use within 1,000 meters of woodpecker colonies in several national forests.[27] Environmental organizations in the Pacific Northwest may succeed in obtaining broader—and economically more devastating—protection for declining spotted owl populations from old-growth timber harvests.[28] A court in early 1989 ordered the FWS to reinstitute listing proceedings,[29] which threw the BLM and Forest Service programmed timber harvests into an uproar and provoked a spirited struggle in Congress (*Public Land News*, Sept. 14, 1989).

The four main commands of the Endangered Species Act—to conserve listed species, to avoid jeopardization, to avoid destruction of critical habitat, and to avoid taking—are now firmly entrenched in public natural resources law. The federal agencies realize that the presence of a listed species dramatically alters the management equation. But federal agencies are political entities, highly responsive to political considerations. Without an informed and committed citizenry, willing to litigate as well as castigate, endangered species protection on federal lands can easily slip back into the "balancing" mode by which developmental land uses always seem to come out ahead.

THE ESA IN LAWRENCE, KANSAS

The ESA has local as well as comprehensive legal consequences. I live in Lawrence, a pleasant campus town in eastern Kansas. Except for occasional complaints about emissions from some local plant, Lawrence seems oblivious to, or above, the environmental disasters so common in other areas of the country. The local citizens are much more concerned about whether to build a second high school than where their trash will be buried. Even into this island of environmental serenity, however, the reality of endangered species law has intruded. Two recent incidents of little ultimate significance tell us a lot about recent changes in law and public attitudes.

The first is the story of Agnes T. (The) Frog. Agnes is the personification of the northern crawfish frog, a species listed as threatened by the state but not by the federal government. The city, county, and state want to build a bypass road—with federal money, of course. The bypass will traverse a wetland south of Lawrence that may be good frog habitat. Some of the local citizens—who probably detest this road more than they love northern crawfish frogs—are incensed. They formed the Committee to Elect a True Amphibian, and Agnes received 1,850 votes in the last county election. Agnes has also generated a losing lawsuit[30] along with considerable pressure on federal decision makers to accommodate frog needs in the still-delayed environmental impact statement on the project.

Agnes no doubt is an admirable creature in the eyes of her creator. Regrettably, however, no one is able to recount a single contribution she makes to human welfare, even under such self-justifying notions as "ecological health and stability." She is nevertheless legally wonderful. High officials in the Environmental Protection Agency and the highway administration worry about her. So too does the general counsel of the Council on Environmental Quality. Agnes's face will not launch a single ship, but her existence has launched litigation and public outcry. Agnes the Frog is the Joan of Arc to the occupying armies of diggers, blasters, accountants, builders, and speculators—our rallying point against the kind of material progress that destroys trees, bogs, and frogs. She forces us to realize that every benefit has a cost. The truly amazing thing is that so many people are unwilling, and vehemently so, to pay the cost of one frog for one new road.

After the road is built, as it will be, memories of Agnes will fade quickly. Perhaps her kin will become plentiful elsewhere. Perhaps the northern crawfish frog will only be found in natural history museums, embalmed and dissected. I do not know. I do know that the story of Agnes is being repeated all over the country as "insignificant" creatures—spotted owls, snail darters, rare butterflies, red-cockaded woodpeckers, furbish louseworts, bats—become the legal hope for people who have decided that they have enough dams or shopping centers or roads or timber and not enough places where the cursed of man can breed in peace.

The second incident in Lawrence concerns the relative priority of bald eagles and retail outlets. Bald eagles are one of the few unquestionable success stories of federal conservation policy. Since DDT was banned and other forms of killing were more effectively policed under the ESA, the eagle has made a strong comeback. In fact, a population now winters on the Kansas River almost in the heart of Lawrence. The area where the downtown meets the river is a municipal disgrace where ancient, decrepit buildings dominate. To the delight of everyone who matters in Lawrence, some out-of-state outfit proposed to renovate one such building into a shopping mall. The only obstacle to the project was the requirement that the developer obtain a permit from the Army Corps of Engineers for its minor alteration of the riverfront. Enter the eagles.

Lawrence's bald eagles perch in cottonwoods along the river and fish in the ice-free areas by the power plant and the dam. The mall development as originally proposed would have necessitated chopping down cottonwoods on the south riverbank. Again, local controversy erupted. Eagle partisans predicted dire consequences, and the local newspaper harrumphed about the overriding need for economic development. With the aid of the local Audubon people, a compromise was worked out whereby some trees were saved and some restrictions imposed on mall use in the wintering season. A last-minute lawsuit to save the trees was summarily dismissed on procedural grounds.[31] The police brought coffee and doughnuts to the diehards who perched in the tree while the chainsaws waited patiently. The protesters were arrested—very, very gently—and undoubtedly will receive little punishment for their forlorn, peaceful vigil. The trees came down.

The mall, like the road, will be built. Whether the eagles will suffer, or to what extent, cannot be known until the event. The point is that the law forced all concerned to turn their attention from profit to birds, because unless they could devise a plan to minimize the conflict, the birds would win; and the humans knew it.

REFLECTIONS

Reflections on the true meaning and impact of the ESA after more than fifteen years lead to mixed conclusions. On the negative side of the ledger,[32] the act apparently is too little and too late in many cases. Biologically, the California condor was simply too far gone from both genetic and human-related causes. Worse, thousands if not millions of species worldwide are doomed in any event; many will disappear through human-caused habitat destruction without ever having been identified or classified. Economically, endangered species protection is expensive, an inevitable concommittant of remediation instead of prevention. Politically, it is too easy for federal agencies to water down legal requirements in freshets of high-sounding phrases. Legally, the ESA remains a limited remedy in spite of its strictness. It is too narrow, focusing primarily on single species in single situations. It does not kick in until the population level of a species is at the danger point, and politics can obstruct listing even when the danger is clear.

A general federal wildlife protection law or a general "ecosystem" approach may not be a panacea either. But if we are to transform our noble (quixotic?) legislative goal into reality, some broader form of protection seems necessary. Human land use is the main cause of habitat destruction, and habitat destruction is the main cause of wildlife endangerment. The hard choice is unavoidable: Saving wildlife means more restrictions on what we can do to the natural environment.

Conceding that the present ESA mechanism is not perfect, the other side of the coin is that endangered species protection in the United States since 1973 has been far more successful than the act's sponsors could have hoped. Those sponsors evidently did not foresee the range of species to which the act would apply or the kind of land-use conflicts that would arise. In spite of political opposition and administrative disinclination in the 1980s, the list of protected species continues to grow. When agencies ignore or downplay the welfare of listed species, courts in the main have forced them back on the path of legal rectitude. Although the presence of an endangered species by itself rarely blocks a proposed land development, it is now accepted that developers must alter their proposals and mitigate consequences. In an imperfect world, perhaps this is the best that the only self-consciously anthropocentric species can do. It is, in any event, far preferable to the state of affairs before 1973.

Nearly a decade after arriving at the cautious conclusion that began this essay, I had recent occasion to revisit the subject in a less theoretical context. From the viewpoint of species in danger, my conclusion today is more hopeful:

> The keys to the puzzle are mitigation and accommodation. Lawyers should inform developers or user clients that their preferred use is subservient to the listed species' welfare, and that their projects or uses must be designed to obviate the likely harms to protected species that they might otherwise cause. The courts are not prone to outlaw all human uses that might conflict with some endangered wildlife if they are convinced that the agency and promoter have made good-faith efforts to observe the spirit as well as the letter of the ESA. As *Thomas, Conner,* and *Sierra Club v. Lyng* illustrate, however, apparent attempts to circumvent ESA strictures are prescriptions for litigative failure and the delay, expense, and frustration that inevitably ensue.[33]

Notes

I apologize for the overabundance of citations to my own books and articles. This, however, is a personal essay, not an objective evaluation, so I will confine myself to my favorite sources. The reader who is so inclined may consult the more detailed treatments of the subjects briefly alluded to in these citations.

1. In Genesis, God commanded Man to assert dominion "over every thing that moves." This injunction has been observed much more closely than other biblical lessons.

2. 16 U.S.C. §§1531–1543 (1982).

3. 16 U.S.C. §§703–711 (1982).

4. 16 U.S.C. §668 (1982).

5. For example, 16 U.S.C. §§528–531 (1982); 43 U.S.C. §§1701–1784 (1982). See generally Coggins and Wilkinson (1987).

6. See, for example, *Perkins v. Bergland*, 608 F. 2d 803 (9th Cir. 1979); see also Coggins (1982).

7. See Coggins (1990.)

8. I confine my attention to mechanisms for the conservation of endangered and threatened species of wildlife and plants on the federal public lands not because developments on federal lands are more important than elsewhere, but rather because I am more familiar with the disputes in this sphere.

9. For example, *Defenders of Wildlife v. Andrus*, 11 ERC 2098 and 455 F. Supp. 446 (D.D.C. 1978); see Bean (1977, p. 138).

10. See, for example, *Sierra Club v. Lyng*, 694 F. Supp. 1260, 1268 (E.D. Tex. 1988), appeal pending; Wilkinson (1984).

11. *Cabinet Mountains Wilderness v. Peterson*, 685 F. 2d 678 (D.C. Cir. 1982).

12. *Foundation for North American Wild Sheep v. United States*, 681 F. 2d 1172 (9th Cir. 1982).

13. *TVA v. Hill*, 437 U.S. 153 (1978).

14. *Thomas v. Peterson*, 753 F. 2d 754 (9th Cir. 1985).

15. *Connor v. Andrus*, 836 F. 2d 1521 (9th Cir. 1988), cert. denied, 109 S. Ct. 1121 (1989); *Bob Marshall Alliance v. Hodel*, 852 F. 2d 1233 (9th Cir. 1988), cert. denied, 109 S. Ct. 1340 (1989).

16. *Sierra Club v. Marsh*, 816 F. 2d 1376 (9th Cir. 1987); but compare *Robertson v. Methow Valley Citizens Council*, 109 S. Ct. 1835 (1989).

17. *New England Naturalist Ass'n v. Larsen*, 692 F. Supp. 75 (D.R.I. 1988); *Carson–Truckee Water Conservation District v. Clark*, 741 F. 2d 257 (9th Cir. 1984), cert. denied, 105 S. Ct. 1482 (1985).

18. *Wilson v. Block*, 708 F. 2d 735, 747-51 (D.C. Cir.), cert. denied, 464 U.S. 956 (1983); *Enos v. Marsh*, 769 F. 2d 1363 (9th Cir. 1985); *Vance*

v. Block, 635 F. Supp. 163 (D. Mont. 1986), rev'd on other grounds, 840 F. 2d 714 (9th Cir. 1988); *NWF v. National Park Service*, 669 F. Supp. 384 (D. Wyo. 1988).

19. *Sierra Club v. Clark*, 755 F. 2d 608 (8th Cir. 1985).

20. *Defenders of Wildlife v. Andrus*, 428 F. Supp. 167 (D.D.C. 1977).

21. *National Wildlife Federation v. Hodel*, 23 ERC 1089 (E.D. Cal. 1985); *California Fish and Game Comm'n v. Hodel*, 18 ELR 20141 (E.D. Cal. 1987).

22. *Defenders of Wildlife v. Administrator*, 688 F. Supp. 1334 (D. Minn. 1988), aff'd in part, 882 F. 2d 1294 (9th Cir. 1989).

23. But compare *Connor v. Andrus*, 453 F. Supp. 1037 (W.D. Tex. 1978).

24. 12 ERC 1156 (D. Neb. 1978).

25. *Riverside Irrigation District v. Andrews*, 758 F. 2d 508 (10th Cir. 1985). See the essay by Dan Tarlock in Part III of this volume.

26. *Carson–Truckee Water Conservation District v. Clark*, 741 F. 2d 257 (9th Cir. 1984), cert. denied, 105 U.S. 1482 (1985).

27. *Sierra Club v. Lyng*, 694 F. Supp. 1260 (E.D. Tex. 1988), appeal pending.

28. See *Portland Audubon Society v. Hodel*, 866 F. 2d 302 (9th Cir. 1988), cert. denied, 109 S. Ct. 3229 (1989), on remand, 884 F. 2d 1233 (9th Cir. 1989).

29. In *Northern Spotted Owl v. Hodel*, 716 F. Supp. 479 (D. Wash. 1988), the court held that the FWS acted arbitrarily in refusing to list the owl as endangered or threatened in light of the available scientific evidence of the species' status. See *Public Land News*, March 30, 1989, p. 1.

30. The case, currently on appeal, involves primarily questions of municipal finance.

31. *Protect Our Eagles' Trees (Poets) v. City of Lawrence*, 715 F. Supp. 996 (D. Kans. 1989).

32. I can never remember whether that is the credit or debit side.

33. See Coggins (1990, § 15.05 [4]).

References

Bean, Michael. 1977. *The Evolution of National Wildlife Law*. New York: Praeger.

Chase, Alston. 1986. *Playing God in Yellowstone*. Boston: Atlantic Monthly Press.

Coggins, George C. 1975. "Conserving Wildlife Resources: An Overview of the Endangered Species Act of 1973." *N.D.L. Review* 51:315.

Coggins, George C. 1982. "Of Succotash Syndromes and Vacuous Platitudes: The Meaning of 'Multiple Use, Sustained Yield' in Public Land Management." *University of Colorado Law Review* 53:229.

Coggins, George C. 1990. *Public Natural Resources Law*. New York: Boardman.

Coggins, George C., and Parthenia Evan. 1982. "Predators' Rights and American Wildlife Law." *Arizona Law Review* 21:821.

Coggins, George C., and Margaret Lindeberg-Johnson. 1982. "The Law of Public Rangeland Management II: The Commons and the Taylor Act." *Environmental Law* 13:1.

Coggins, George C., and Irma S. Russell. 1982. "Beyond Shooting Snail Darters in Pork Barrels: Endangered Species and Land Use in America." *Georgetown Law Journal* 70:1433.

Coggins, George C., and Charles Wilkinson. 1987. *Federal Public Land and Resources Law*. 2nd ed. Mineola, N.Y.: Foundation Press.

Lund, T. 1980. *American Wildlife Law*. Berkeley: University of California Press.

Public Land Law News. March 30, 1989, p. 1.

Public Land Law News. Sept. 14, 1989, p. 1.

Reed, N., and D. Drabelle. 1987. *The Fish and Wildlife Service*. Boulder: Westview Press.

Swanson, Gustav A. 1978. "Wildlife on Public Lands." P. 428 in H. Brokaw (ed.), *Wildlife and America*. Washington, D.C.: Council on Environmental Quality.

Trefethen, James B. 1975. *An American Crusade for Wildlife*. New York: Winchester Press.

Wilkinson, Charles. 1984. "The Forest Service: A Return to First Principles." *Public Land Law Review* 5:1.

PART II

COMPONENTS OF THE FEDERAL ENDANGERED SPECIES PROGRAM

The wild things of this earth are not ours to do with as we please. They have been given to us in trust, and we must account for them to the generations which will come after us and audit our accounts.

—WILLIAM T. HORNADAY

And of every living thing of all flesh, two of every sort shalt thou bring into the ark.

—GENESIS

THE ENDANGERED SPECIES LISTS: CHRONICLES OF EXTINCTION?

by

WILLIAM REFFALT

For one species to mourn the death of another is a new thing under the sun. The Cro-Magnon who slew the last mammoth thought only of steaks. The sportsman who shot the last pigeon thought only of his prowess. The sailor who clubbed the last auk thought of nothing at all. But we, who have lost our pigeons, mourn the loss. Had the funeral been ours, the pigeons would hardly have mourned us. In this fact, rather than in Mr. DuPont's nylons or Mr. Vannevar Bush's bombs, lies objective evidence of our superiority over the beasts.

W HEN ALDO LEOPOLD PENNED "On a Monument to the Pigeon" in 1947, he conveyed a note of optimism. Now, forty-one years later, there are signs that his optimism was premature.

Nearly seventy-four years have passed since the carefully recorded demise of the last passenger pigeon, a twenty-nine-year-old female residing in the Cincinnati Zoological Gardens. That same month, September 1914, chronicled the passing of the last captive Carolina parakeet as well. Their quiet deaths echoed the earlier extinction of the Labrador duck and great auk, reawakening the nineteenth-century wave of concern inspired by the near extinction of the bison and the ominously increasing documentation of drastic declines in many other species. Yet even as America realized that its forests would never again feel the hurricane might of the pigeon profusion or witness the bright chattering flocks of parakeets, committed men and women were using that tragedy to generate support for increasing efforts to stem the tide of extinctions that appeared ready to engulf America's prominent wildlife species.

During that era there were no official lists of endangered wildlife. A few books and magazine articles presented terse descriptions of vanished species and listed those deemed likely to follow the pigeon and several other members of our wildlife heritage to the abyss of extinction. (See, for example, Hornaday, 1913.)

THE OFFICIAL LISTS

When the first Endangered Species Act (ESA) was passed in 1966,[1] the secretary of the interior was directed, after consultation with affected states and scientific groups having expertise in the matter, to "publish in the Federal Register the names of the species of native fish and wildlife found to be threatened with extinction." This congressional directive to create the first official U.S. endangered species list postdated the July 1966 Red Data Book, published by the International Union for the Conservation of Nature, containing lists of rare and endangered species around the world. Significantly, when the first ESA passed, the expressed intent was to protect thirty-five or so species of native mammals and thirty to forty species of native birds believed to be near extinction (*Congressional Quarterly Almanac*, 1966).

Through the 1969 Endangered Species Protection Act,[2] Congress institutionalized the global scope of the listing process by authorizing the interior secretary to list wildlife "threatened with worldwide extinction" and to prohibit the importation of such species into the United States except under specified conditions. The decision to list a species required consultation with the states that might be affected and, in the case of foreign species, with the nations where the species is normally found. In contrast to the earlier, somewhat informal process for listing, the 1969 act intended that listing decisions comply with the Administrative Procedures Act (Bean, 1983, p. 323).

With the enactment of Public Law 93-205 on December 28, 1973, and subsequent amendments, the ESA became the most far-reaching wildlife statute ever adopted by any nation. Under the 1973 act, the listing process became more demanding and even more formalized. With the addition of plants and the expansion of the term "fish or wildlife" to include all animals (vertebrate as well as invertebrate), many additional species became subject to listing. In 1978, Congress further required that the listing of a species be accompanied by the designation of critical habitat—an often difficult task even for well-known but highly mobile wildlife. The legislation even established procedures and time limits for acting on listing petitions. During the previous seven years, 109 U.S. species and 300 foreign species had been listed (*Congressional Quarterly Almanac*, 1973). Following passage of the 1973 act, 346 species were listed over the next seven years (Di Silvestro, 1985).

CURRENT LISTING STATUS

As of May 1988, a summary of ESA listings and recovery plans produces the following approximate numbers of species, subspecies, or populations:[3]

- U.S. endangered: 377
- U.S. threatened: 118
- Total U.S. listed: 495
- Approved recovery plans: 229 (covering 283 species)
- Foreign endangered: 467
- Foreign threatened: 39
- Total foreign listed: 506

Thus nearly twenty-two years after the first ESA and more than fifteen years after enactment in its current form only 57 percent of U.S. listed species have approved recovery plans—a record that fails to demonstrate attainment of the basic promises of the act. When coupled with the listing backlogs, the task of achieving those promises seems herculean. This is what the record shows:

- Approximately 34 percent of likely U.S. threatened or endangered species have been officially listed (that is, listed + Category 1).

- Approximately 11 percent of possibly threatened or endangered species have been officially listed (that is, Category 2 added to the preceding total).
- Approximately 6 percent of possibly threatened or endangered species have approved recovery plans.

Of the 495 U.S. listed species, only about 16 (3.2 percent) are recovering. Another 18 listed species (3.6 percent) may already be extinct. For most species, it is almost impossible to make a positive determination that absolutely no more of their kind live on this planet. The recent discovery of several ivory-billed woodpeckers on Cuba, for example, followed three decades in which no verified sightings were made.

Another list records the "delisted" species. Since the creation of such statistics, the Fish and Wildlife Service (FWS) has logged fifteen delisting actions. In the good news category, three of these delistings have been due to recovery of the species and two (the brown pelican and American alligator) represent recovery over portions of their range. Another four of the delisting actions have been due to corrections of the original data or status. The bad news is that six of these delisting actions (40 percent) resulted from the species becoming officially extinct.

EFFECTS OF LISTING

Several benefits are bestowed upon endangered or threatened species once federal agencies officially list them. Listing conveys greater recognition of a species' precarious status as well as the need for restraint in modifying any conditions affecting them. Agencies, organizations, and individuals must consult with the Fish and Wildlife Service or the National Marine Fisheries Service before any federally assisted action modifies a listed species' habitat or when individuals of the species might be taken. Listing also shields endangered or threatened species by applying restrictions on the taking or trafficking of such species. Finally, listing leads to creation of a species recovery plan and facilitates acquisition of land or interests in land to assist recovery. Such conservation safeguards are not inconsiderable in their effect. Sometimes such measures can halt excessive losses of individuals in the population, thereby arresting a population decline and permitting adequate time to take recovery actions— or, in some cases, allowing the species itself to recover its numbers. Where the threshold of adequate recruitment remains unmet, however, or habitat has been rendered incapable of supplying vital functions, careful measures are essential to recovery. For species on the official list, a great deal can happen before the recovery actions are outlined in a plan, the plan is approved, and its details are successfully implemented. For endangered but unlisted species, serendipity comprises the most likely salvation.

THE BACKLOG PROBLEM

In October 1975, less than two years after passage of the expanded act, the Fish and Wildlife Service testified that it had received petitions to list 23,962 species in addition to 144 listing actions undertaken on its own authority. Such backlogs have had a huge impact on implementation of the ESA. To cope with the backlogs, the FWS developed a triage system in which priority was given to the higher orders of animals and species that agency officials believed would respond best to recovery actions (Schreiner, 1975, p. 16). Yet this approach has produced heated controversy. And neither the issues nor the backlogs have vanished.

Over the years, the FWS has compiled several prelisting lists. The candidate Category 1 list (now totaling about 950 U.S. species) includes those species, subspecies, or populations for which the data support official listing as either threatened or endangered but which have not been listed due to lack of a priority or inadequate resources to complete the formal listing process. It is currently believed that about 177 (18.6 percent) of the Category 1 entries already may be extinct. Candidate Category 2 includes species, subspecies, or populations for which the data are considered inadequate to make a listing decision but some source has provided evidence that the status of the species is either threatened or endangered. This list includes approximately 2,944 entries; about 118 (4 percent) of Category 2 may be extinct.[4]

ASSESSMENT OF THE PAST

When the first evidence of severe wildlife depletion was compiled over a century ago, it aroused actions by the individual states and several federal agencies. The agencies' actions ultimately achieved a good measure of success. Elk, deer, antelope, turkeys, and several other game species responded well to management actions including stricter hunting limitations (especially termination of market hunting), introduction of individual animals into former, recovered ranges, and establishment of parks, forests, refuges, and other protected areas. Another factor aiding recovery of several species was the initial, fortuitous recovery of forest habitats after unchecked logging had denuded whole landscapes as manifest destiny pushed America's frontier to California's rugged coastline. Human populations were sparse at this time, and demands on wildlife and wildlands were moderate by today's standards. The species in immediate jeopardy were also those readily granted public sympathy, and agency budgets were adequate to accomplish the common management measures then known to biologists.

With no official endangered species lists, federal agencies were largely able to develop programs at the pace and in the priority order they chose. There were no unyielding public interest groups pressing agencies to hold hearings or write impact statements and no second-guessing by consulting biologists under contract to corporate interests. There was a limited, well-known, and popular group of animals to attend to, and as their needs became known and were met these (mostly game) species responded favorably. The largest interest group, hunters, provided immediate praise and bigger license sales that provided more resources to expand the cycle of success. Other conservationists also praised early wildlife conservation efforts. They spread the words of success and desiderata to Congress and millions of Americans through their testimony, books, magazines, meetings, and films. Both federal and state budgets benefited.

By 1966, however, when the ESA was deemed necessary, things had become more complex, human populations had redoubled, and more and more land was being used for maximum benefit to the "bottom line." Numerous little-known species were being evicted for "higher and better" uses of land and water. As the ESA developed and its scope broadened, many "ugly ducklings" and species whose needs sometimes conflicted with development plans became candidates for listing. Biologists often knew very little about the requirements of such species or the full extent of their habitats. Funding for an enlarged endangered species program was hard to come by, and the popular management and research programs left little resources available for these nondescript and often controversial creatures.

Requirements for listing became more precise and more demanding, and the potential candidates for federal protection grew into the thousands. The public became more interested, more involved, and more demanding. Some asked, "What good is that creature anyway?" Answers were not easily at hand and often sounded fanciful. Agencies quickly found that less controversy meant less discomfort. Since there was more to be done than money or time allowed, priorities had to be set. Hence the prevailing attitude became: "Why not do the less controversial ones first?" The listing process has thus become a focus of attention and a means of controlling the pace of the entire ESA effort. In the past seven years, only 246 species have been listed—40 percent fewer than in the first seven years and 29 percent fewer than in the second seven years since passage of the 1966 act.

By the time species qualify for official listing, their ecological situation is frequently critical and the options (if known) for their recovery are quite limited. Delayed listing often results in high research and management costs. Thus agencies are constantly faced with balancing high costs and high-risk efforts (those that include data gaps and great uncertainty) against those requiring fewer resources and having a greater likelihood of success. Endangered species work is not a low-tension job.

Given such a background and perspective, two fundamental questions arise: Can the inexorable rise of extinctions be averted? And if so, how?

WHAT THEN THE FUTURE?

A deep chesty bawl echoes from rimrock to rimrock, rolls down the mountain, and fades into the far blackness of the night. It is an outburst of wild defiant sorrow, and of contempt for all the adversities of the world.

Every living thing (and perhaps many a dead one as well) pays heed to that call. To the deer it is a reminder of the way of all flesh, to the pine a forecast of midnight scuffles and of blood upon the snow, to the coyote a promise of gleanings to come, to the cowman a threat of red ink at the bank, to the hunter a challenge of fang against the bullet. Yet, behind these obvious and immediate hopes and fears there lies a deeper meaning, known only to the mountain itself. Only the mountain has lived long enough to listen objectively to the howl of the wolf. [Leopold, 1949, p. 129]

Americans have taken bold steps, through the ESA, to sustain the remarkable wildlife heritage of this nation. But becoming the responsible beings urged by Aldo Leopold in "Thinking Like a Mountain" calls upon us to make even stronger commitments. The most recent amendments to the ESA passed in 1988 contain some of what is called for: monitoring programs for candidate species, strengthened plant protection, increases in authorized funds, manpower to accelerate all facets of the ESA efforts, and greater penalties for those who would extinguish a living spark forever and destroy a piece of our common heritage.

But even these sorely needed thrusts fall short of the monumental task at hand. Perhaps the time has come for Congress to emulate the wisdom and objectivity of the mountain and to set aside its tendency to view each potential conflict over a species recovery effort with a legislator's microscope. Perhaps it is time to summon the foresight and courage to legislatively place all candidate Category 1 species on the official list without further ado. Congress also needs to provide new direction for identifying, listing, and implementing plans to recover endangered wildland communities—those in which several species are threatened because their basic habitat has become drastically modified or fragmented and is unable to sustain itself unless protective steps are taken soon.

Too often, listing species has been akin to giving morphine to cure a hemorrhage. The time has come to do more than ease the pain. We need to stabilize endangered habitats while completing the analysis necessary to take curative steps for full recovery. Human judgments of the patient's appearance or net economic worth impede action. Only mountain-like objectivity and wisdom will cure these patients. For one species to take great pains to avert the death of another is the "new thing under the sun" this generation should impart to the future.

Notes

1. Pub. L. 89-669, 80 Stat. 926.
2. Pub. L. 91-135, 83 Stat. 275.
3. Statistics compiled by the Wilderness Society from FWS data, June 1988.
4. The data relating to Category 1 and 2 species are from the 1986 and 1987 Defenders of Wildlife reports, *Saving Endangered Species*. They were verified with FWS officials by the Wilderness Society in June 1988.

References

Bean, Michael. 1983. *The Evolution of National Wildlife Law*. Rev. ed. New York: Praeger.

Di Silvestro, Roger L. (ed.). 1985. *Audubon Wildlife Report*. New York: National Audubon Society.

Fisher, James, N. Simon, and J. Vincent. 1969. *Wildlife in Danger*. New York: Viking.

Hornaday, William T. 1913. *Our Vanishing Wildlife*. New York: New York Zoological Society.

Leopold, Aldo. 1949. *A Sand County Almanac*. New York: Oxford University Press.

Schreiner, Keith. 1975. *Hearing Record: Endangered Species Oversight*. House Subcommittee: Fish and Wildlife Conservation and Environment, October 1, 1975, Serial No. 94-17. Washington, D.C.: Government Printing Office.

Terres, John K. 1980. *The Audubon Society Encyclopedia of North American Birds*. New York: Knopf.

U.S. Fish and Wildlife Service. 1981. *Endangered Means There's Still Time*. Washington, D.C.: Government Printing Office.

AVOIDING ENDANGERED SPECIES/ DEVELOPMENT CONFLICTS THROUGH INTERAGENCY CONSULTATION

by

STEVEN L. YAFFEE

SOMETIME IN EARLY 1973, in the organized chaos of the Ninety-third Congress, a sleeper was added to a developing piece of environmental legislation. Preservationists, administration experts, and the House Merchant Marine and Fisheries Committee staff pushed for including in the Endangered Species Act (ESA) an absolute mandate for federal agencies to protect endangered species. Where previous legislation had required federal agencies to provide protection "where practicable," Section 7 requires agencies to take "such action necessary to insure that actions authorized, funded, or carried out by them do not jeopardize the continued existence of an endangered species." Further, it requires the agencies not to destroy critical habitat and prescribes a mechanism for implementing the requirement via interagency consultation.

The change from the old consultation requirement to the new mandate was significant. While the concept of interagency cooperation through consultation was well defined in wildlife law by the early 1970s (Bean, 1977, p. 192), it had yielded little substantive effect. Just as my pleas to my young daughter to "cooperate" fall on deaf ears when she does not think cooperation is in her best interests, so do congressional mandates that ask one agency to help another. The old statute required action to protect endangered species "insofar as is practicable and consistent with [the agency's] primary purposes." By 1973, history, to say nothing of common sense, suggested that the odds of getting the U.S. Department of Transportation to alter plans for an interstate highway to protect a rare bird, or the Army Corps of Engineers to forgo a dam to protect a population of endangered bats, were slim indeed. Then along came Section 7.

THE WISDOM OF SECTION 7

One of the remarkable things about Section 7 of the Endangered Species Act is that it has survived relatively intact. Although the interagency consultation process has been refined, streamlined, and provided with an almost-never-used exemption process, the mandate remains solid. It survives because it makes sense substantively, because it has been implemented flexibly and adaptively (perhaps too much at times), and because environmentalists have mustered enough support in Washington to turn away major challenges. Establishment of the so-called God Committee,[1] the high-level exemption-granting group, might have been feared by the environmentalists in 1978, but it has provided an essential political pressure valve: You say you have an irreconcilable conflict? Well, we can handle that. Put it to the test.

While Section 7 is problematic as development policy, as endangered species policy it makes sense. It recognizes the fact that a regulatory program is only as good as its implementation network. Public policy functions by changing the behavior of a range of parties. In the case of endangered species policy, action must come not only from the Fish and Wildlife Service (FWS) and the National Marine Fisheries Service, but from a range of public and private bodies as well. Statutes create the conditions for such networks of institutions to function in two ways: by enabling action where there is mutual interest and by structuring incentives to encourage groups to act appropriately. The Endangered Species Act uses both of these mechanisms. Section 6 grants, which provide financial assistance to state endangered species programs, allow for state/federal cooperation where mutual interest exists. Interagency cooperation, as defined by Section 7, influences the incentives that agencies face.

Section 7 also works fairly well because it recognizes the strategic disadvantage that nonhuman (and nonvoting) lifeforms have. While the absolute mandate rarely results in absolute action, it has raised the priority of endangered species preservation on the agenda of numerous federal development agencies by making it disadvantageous not to do so. Ever since the Supreme Court's Tellico Dam ruling that Congress intended the absoluteness of the mandate contained in Section 7,[2] development agencies have recognized the potential for delay and controversy that can result from failing to consider endangered species (at least somewhat) in planning their projects. Similarly, the Fish and Wildlife Service fears the political controversy that can result from visible issues that appear to pit preservation against economic development. While the mandate has created an additional lever that can be used by nongovernmental intervenors, such groups have used the lever sparingly fearing a congressional backlash against the program. As a result, all three groups have the incentive to seek creative solutions that accommodate preservation and development concerns.

The interagency consultation provisions of the ESA have also been significant in expanding endangered species policy into the business of protecting habitat rather than simply protecting individual plants or animals. Since the long-term solution to the endangered species problem requires ecosystem-level planning (U.S. Congress, Office of Technology Assessment, 1987, p. 90) and greater up-front care in land development and management, it makes a lot of sense to incorporate endangered species considerations in ongoing federal planning efforts. Interagency consultation has gotten the endangered species program into land-use planning through the back door (Coggins and Russell, 1982). Indeed, endangered species hot spots (that is, places that witness a lot of controversy over proposed developments) may be fairly reliable indicators of broader planning problems. Hence the prevalence of conflicts between endangered species and proposed developments on a number of western water systems, such as the Upper Colorado and the North Platte rivers, bespeaks a serious resource allocation problem that goes beyond endangered species. Places that evidence a fair number of conflicts over endangered species are in need of deliberate and creative planning. Such planning should consider resource shortages in a manner that balances the variety of human uses while protecting biological systems including endangered species habitat.

THE IMPLEMENTATION RECORD

On the surface at least, the record of the implementation of Section 7 suggests that such planning is possible. While the total number of consultations carried out by the FWS has increased dramatically, the number of projects that have been delayed significantly or stopped has been remarkably small. For example, the total number of consultations more than quadrupled in the seven-year period between 1979 and 1986 (Alderson, 1987, p. 15). The number of jeopardy opinions—cases in which the FWS determined that a project was likely to harm an endangered or threatened species—was consistently less than 1 percent of all consultations (Bean, 1986, p. 357).

Fish and Wildlife Service documentation suggests that very few of these projects were canceled due to an endangered species conflict. Of the 86 projects that received jeopardy opinions in fiscal years 1982, 1983, and 1984 (out of 18,670 consultations including 922 formal consultations), only 14 were canceled or withdrawn—and only a portion of these for reasons of endangered species protection (U.S. Fish and Wildlife Service, Office of Endangered Species, undated). More typically, mitigating measures are designed into the projects to offset or avoid threats to a species. For example, developers of a landfill in San Jose, California, agreed to a number of mitigation measures to offset the project's impact on the Bay checkerspot butterfly—creation of a conservation trust fund, ongoing research, habitat acquisition and management, restoration and revegetation when the landfill is at capacity, and off-site reintroduction and recovery (Murphy and Freas, 1988). Similarly, biological opinions on several marina projects on the intracoastal waterway in Florida have prescribed a variety of measures to offset impact on the West Indian manatee, including reducing the number of power boat slips, reducing boat speed limits, conditioning slip rental agreements, and informing boat owners about the threat to the manatee.

Contrary to common complaints, consultations do not appear to result in excessive delay of projects. The 922 formal consultations carried out in fiscal years 1982 to 1984 averaged fifty days to complete. Consultations that received jeopardy opinions, and hence involved the most work, were completed on average in three months. Since endangered species considerations are simply one aspect of most federal permitting or development processes, three months of review time is most likely a drop in the project review bucket—an insignificant component if consultation is initiated early in the regulatory process.

An analysis of conflicts between the consultation requirements of the Endangered Species Act and western water projects provides the best "worst case" data. In 1987 the General Accounting Office analyzed the effects of the consultation requirements on water projects in seventeen western states and concluded:

> Consultations carried out under the Endangered Species Act have had little effect on western water projects. While sixty-eight consultations affected projects over the seven and a half year period we examined, for the most part these effects have not been major. Further, even when the consultation affected the project, Department of the Interior and other agency officials indicated that other events occurring at the same time (such as difficulties in arranging project financing) sometimes had a more significant effect than the consultation process. The willingness and ability of the Service and project sponsors to arrive at compromise solutions when conflicts occurred also contributed to reducing the consultation requirement's ultimate effect on project development. [U.S. General Accounting Office, 1987, p. 30]

While the implementation record is reassuring to those who are concerned that the ESA has delayed development or increased its cost, the same data can give rise to concerns that species might be receiving inadequate protection. Does a low number of projects canceled due to conflicts with endangered species mean that species are not receiving enough consideration? Does a low number of jeopardy opinions mean that the conflicts have been worked out—or that the FWS is trying to minimize controversy, perhaps at some expense to endangered species? Does the fact that the FWS increasingly has used informal consultation mean that the agencies are getting better at designing projects considering the needs of endangered species—or that the FWS is trying to keep the process out of the more costly and potentially more controversial terrain of formal consultations? (Formal consultations as a percentage of total consultations declined from 38 percent in 1979 to roughly 4 percent in 1989.) While there are no clear answers to these questions, I suspect that the effect of the Section 7 consultation process varies from agency to agency and from project to project. For most agency decisions, Section 7 has ensured that endangered species are at least considered as part of the impact assessment; for a number of projects, consultation has resulted in significant changes in project design; and for a few projects that seemingly threaten a major and unmitigable impact on a species, the Section 7 process has created an additional window into agency decision making, by providing information and opportunity for proponents of endangered species protection to influence federal permit and development decisions.

PROBLEMS IN IMPLEMENTATION

Although there have been problems with implementation of the consultation process, these problems have been largely with the implementation of the statute and not the law itself (if the distinction can be made). As with other elements of the ESA, consistently limited funding has constrained FWS activity in the consultation process. Over the past decade, consultation activity has more than quadrupled, yet budgets and staff allocations have stayed constant at roughly $2.5 million per year. Either FWS personnel have gotten four times more efficient at carrying out consultations (an unlikely situation) or the amount of staff effort per consultation has declined. The shift from formal to informal consultation no doubt reflects the workload problem: Informal consultations take less time, provide more room for negotiation, and avoid the bureaucracy of formal consultation.

One should also recognize that participants in the consultation process follow their incentives, many of which run counter to the interests of endangered species protection. No doubt the FWS has begun using informal consultations in part to avoid the potential for controversy that formal consultation entails. Likewise development agencies face a disincentive to volunteer information that will establish a threat to a listed species, and legislators and nongovernmental parties of all kinds participate in the process in whatever way they can to influence its outcome.

Political considerations clearly are present in the consultation process. Indeed, it would be surprising if they were not. Consultation decisions are allocation decisions. They contribute to an on-the-ground definition of who gets what resources and when. The December 1986 decision on the Stacy Dam in west central Texas is a decision that allocates water resources and associated habitat to a variety of interests in the region, including those of the Concho water snake (*Endangered Species Technical Bulletin*, Feb. 1986 and Jan. 1987). One should expect the Texas congressional delegation to get involved in the issue—as they did. And one should expect the FWS to face a strong incentive to respond to such pressure—as they did in accepting a mitigation plan for the snake that they had rejected in May of the same year as unlikely to work (Alderson, 1987, p. 15). Rather than bemoaning the fact that politics enters into endangered species decision making, we should recognize the realities of the situation and work to exploit the benefits of political inputs—as sources of information about collective values and how intensely they are held—and minimize the negative effects of such forces on species preservation.

IMPROVING THE PROCESS

How do we walk this tightrope? If I could design the ideal consultation process, it would include the following features: It would be absolutely firm on ends—preservation of species—yet flexible on means to reach those ends; it would provide the resources, including staff and information, necessary to implement the policy effectively; it would address the magnitude of the underlying problem; and it would be proactive, not limited to crisis-oriented management.

Section 7 advances us toward this type of process, but we need to go further. As we have seen, the absolute mandate provided in the act, as well as the opportunities provided for judicial intervention by proponents of preservation, help to create an incentive structure that leads to the first objective. Yet we can go further. We need to ensure that the process remains visible by providing information on informal and formal consultations in a format that makes it accessible to nongovernmental groups that can serve as watchdogs. A sample of projects on which consultation has taken place should be examined periodically (preferably by an uninvolved party) to evaluate the actual consequences to habitat and hence the adequacy of current consultations. To promote creativity in devising compromise solutions, training is needed for FWS staff and development agency personnel to ensure that the full range of mitigation techniques are used and that development options are fully explored. In some situations, this inventing of options might require FWS biologists to learn more about development alternatives than they otherwise might wish.

We want to promote consensus in dealing with these conflicts, but not at the cost of long-term harm to an endangered species. The Windy Gap strategy probably goes too far. Named for a project on the Colorado River in north central Colorado, the Windy Gap approach signaled the influence of the Reagan administration on endangered species decision making. Up to 1981, the FWS held that any water diversion from the Upper Colorado would probably jeopardize resident endangered fish and argued that more data were needed to evaluate fully the effects of a diversion. Beginning with the Windy Gap project and continuing for more than thirty additional projects, the FWS began issuing "no-jeopardy" opinions, allowing project construction to proceed contingent on project sponsors contributing to a fund that would pay for impact studies and conservation measures. Environmentalists noted that if the impact studies found the depletions were deleterious, it would be too late. Levying fees as part of a mitigation package is not in itself objectionable; the Grayrocks Dam project on the North Platte used this as a component of a mitigation package to protect a whooping crane stopover point. But simply allowing projects to proceed if they pay tribute without knowing what will happen is wrong.

A consistent set of resources is needed to ensure that the consultation process works as it should. These funds should include an increase in budget and staff slots for the FWS (and probably for their counterparts in development and permitting agencies). Besides the training noted earlier, the FWS (and National Marine Fisheries Service) staff should be trained in negotiation skills. After all, their long-term success in what is almost always a negotiation process lies in their power to persuade, develop alternatives, and mobilize support for their position.

Moreover, information is needed to make the consultation process work better. While certainly more research into the needs and dynamics of endangered species populations would help resolve some of the technical uncertainty prevalent in consultations, better notification processes would be helpful to disseminate information on the existence and habitat needs of endangered species. Developing computerized sources of endangered species information—and making this information accessible to project developers and environmental groups (with appropriate safeguards to avoid publishing the location of sensitive species that could be "collected" to death)—might help to avoid some of the conflicts before they arise. Proponents of development often prefer a tough yet certain situation to a less constrained but ambiguous one. Part of the problem with the consultation process from a developer's standpoint is its mystery and uncertainty. Some of this ambiguity is due to the lack of information on many species. But to the extent that we can devise information systems that allow developers to avoid problems, we will be better off.

An effective consultation process should also deal with the magnitude of the endangered species problem. It is here that action is particularly needed. Consultations should include any species that might be jeopardized by a proposed development, not just those that have made it through the listing gauntlet. Impacts on candidate species should be evaluated as part of the informal consultation process, even though the ESA's absolute mandate to protect will not (and perhaps should not) apply. While the regulatory process of necessity functions on a case-by-case basis, more work on evaluating and forecasting cumulative effects of development is needed.

Finally, the consultation process should not be limited to projects on U.S. soil. Activities that involve federal government support or permits overseas should be evaluated through inter-agency consultation. Although the act provides for foreign consultations, implementation has fallen far short of its potential.[3] Since a good deal of the global endangered species problem lies in the loss of habitat in the tropics, any lever that we have to stem the tide of extinction should be used. Section 7 constitutes a regulatory process that could have an effect.

The approach to endangered species protection established by the ESA is fundamentally a reactive process. It responds to crisis, proposed developments, and alleged endangerment. The persistent poverty that has attended the program has limited any attempts to be proactive through strategic planning for future developments that avoid species conflicts. The ESA provides a good stop-gap approach to species conservation, and the consultation process moves us in the right direction. But the long-term solution must come from more effective planning at all levels of administrative organization. This is especially true as we broaden our view of endangered species preservation to include the ecosystems and processes of which such species are part and that give them vitality and meaning.

The endangered species problem is principally a land-use problem, and it will only be solved through deliberate action that utilizes expertise and involves all affected parties in decision making. How do we achieve this type of collective decision making in a society that tends to be suspicious of planned change? We do it by using existing federal and state planning processes, such as those for Forest Service and BLM lands, more effectively, by breathing life into moribund regional planning institutions such as river basin commissions, and by creating ad hoc groups, especially around endangered species hot spots. The endangered species coordinating committees for the Upper Colorado and the North Platte rivers are steps in the right direction. Regional biodiversity councils formed by interagency memorandums of understanding or by new legislation might also help provide direction.

Improving the consultation process—and, indeed, improving other elements of the endangered species program—demands leadership. It requires a federal administration cognizant of the value of biological diversity and committed to endangered species preservation. It demands a lead administrative agency that is given a chance to succeed. Both of these items require constituency building. Endangered plants and animals will only be protected as long as there is the political will to do so. A policy is only as good as its support coalition. Proponents of protection must continue to lobby and teach the importance of the issue throughout the populace. Not only does constituent support allow decision makers to make choices that benefit endangered species; it can influence decision makers who otherwise would not care about endangered species.

Implementation requires action from a variety of participants. Government regulation will not solve the problem alone, and the federal budget deficit is certain to limit future federal programs. Expertise and understanding should be cultivated in state agencies. Nongovernmental organizations must continue to play a major role, too, through monitoring, lobbying, educating, and fundraising.

The endangered species problem is a problem of human civilization. It should not be surprising, therefore, that its solution demands changes in our decision-making institutions. Perhaps more than other elements of the ESA, the Section 7 consultation process provides a focal point for the diverse set of human demands that are placed on natural resources. Mediating between these human demands in a way that protects the nonhuman world is a difficult task, but it is essential.

Notes

1. Responding to political pressures to weaken the absolute mandate contained in the Endangered Species Act, amendments enacted in 1978 contained a process by which a project that presented an irreconcilable conflict with an endangered species could be exempted from the requirements of the act after review by a high-level interagency committee. Composed of the secretaries of agriculture, defense, and interior, the chairman of the Council of Economic Advisors, the administrators of the Environmental Protection Agency and the National Oceanic and Atmospheric Administration, and representatives of affected states, the Endangered Species Committee had the power to choose between a project and a species and hence was immediately called the God Committee. In fact, the process established by the amendments has been employed only twice: for the Tellico Dam and Grayrocks Dam projects. Neither received an exemption from the act's provisions.

2. *TVA v. Hill*, 437 U.S. 153 (1978).

3. See, for example, the essay on international components of the ESA by Mark Trexler and Laura Kosloff in Part II of this volume.

References

Alderson, George. 1987. *Saving Endangered Species: Implementation of the Endangered Species Act*. Washington, D.C.: Defenders of Wildlife.

Bean, Michael J. 1977. *The Evolution of National Wildlife Law: A Report to the Council on Environmental Quality*. Washington, D.C.: Government Printing Office.

Bean, Michael J. 1986. "The Endangered Species Program." Pp. 347–371 in Roger L. Di Silvestro (ed.), *The Audubon Wildlife Report 1986*. Orlando: Academic Press.

Coggins, George C., and Irma S. Russell. 1982. "Beyond Shooting Snail Darters in Pork Barrels: Endangered Species and Land Use in America." *Georgetown Law Journal* 70: 1433.

Murphy, Dennis D., and Kathy E. Freas. 1988. "Using the Endangered Species Act to Resolve Conflict Between Habitat Protection and Resource Development." *Endangered Species Update* 5(2–3):6.

U.S. Congress, Office of Technology Assessment. 1987. *Technologies to Maintain Biological Diversity*. OTA-F-330. Washington, D.C.: Government Printing Office.

U.S. Fish and Wildlife Service, Office of Endangered Species. Undated. *Reauthorization of the Endangered Species Act, Section 7 Consultation Summaries, FY1982–1984*. Washington, D.C.: Government Printing Office.

U.S. General Accounting Office. 1987. *Endangered Species: Limited Effect of Consultation Requirements on Western Water Projects*. GAO/RCED-87-78. Washington, D.C.: Government Printing Office.

Yaffee, Steven L. 1982. *Prohibitive Policy: Implementing the Federal Endangered Species Act*. Cambridge: MIT Press.

FEDERALISM AND THE ACT

by

JOHN P. ERNST

Eᴅ Howᴇʟʟ, one of the West's most infamous poachers, personally killed as many as eighty buffalo—approximately one-fourth of the buffalo then in existence. Like many other poachers in the early 1880s, Howell violated state laws with impudence because no one knew if state laws applied in the newly created Yellowstone National Park.

Ed Howell was finally arrested in March 1894—caught red-handed while skinning five freshly killed buffalo. He was soon set free, however. Without a law, he could not be prosecuted. Howell spoke cockily to the press, and the coverage of his case was complete with photos of bloated, decomposing buffalo. The ensuing public outrage helped provide Congress with the incentive to pass the Yellowstone Park Protection Act in 1894, which prohibited hunting in the park.

The Yellowstone Act took years to pass—not because anyone supported poaching, but because those wanting to exploit the park's timber, mineral, and wildlife resources argued that the park should be left to state control. Ninety-two years after Howell killed those five famous buffalo, Yellowstone National Park remains a center of controversy between state and federal wildlife managers, and states' rights is still a battle cry. The buffalo focused attention on state/federal tensions in managing Yellowstone's wildlife in the 1800s, but that role has now been taken over by another animal. Today it is the gray wolf that sits at the center of tensions between state and federal wildlife managers. Moreover, discussing the prospect of reintroducing the wolf into Yellowstone also means discussing the grizzly bear. For the issues surrounding reintroduction of wolves stem from problems governing current management of grizzly bears.

State and federal agencies working together have made notable progress in the battle to reduce the loss of species. The peregrine falcon is soaring high in many states today, and red wolves are roaming the Alligator River National Wildlife Refuge in North Carolina. But the picture is not entirely rosy. State/federal relations have been strained by unreliable funding under Section 6 of the Endangered Species Act (ESA) on the one hand and controversy over the act's strong taking provisions on the other. The almost total restriction on killing or harming a species listed under the act is both a strength and a source of bitter conflict between state and federal wildlife managers.

Many of the current issues relating to federalism and the Endangered Species Act can be traced to the origins of state wildlife management and the history of federal intervention in that process. That history, along with current implementation issues surrounding the act (particularly those pertaining to the gray wolf and the grizzly bear in Yellowstone National Park), illustrates some of the key problems affecting state/federal relations in managing endangered species.

STATE CONTROL OF WILDLIFE

The beginnings of wildlife law can be traced back to the Roman Empire, feudal Europe, and the Magna Carta. At the time of the American revolution, the king and Parliament exercised complete control over wildlife, which they held in the public trust. Following the revolution, the issue was first addressed by the U.S. Supreme Court in 1842. In a landmark decision, *Martin v. Wadell*,[1] the court effectively entrusted the state governments with authority to regulate wildlife once held by the king. That decision served as the basis for the state ownership doctrine that guided wildlife management throughout the nineteenth century.

Concerns about declining wildlife populations resulted in the formation of state wildlife agencies in the 1800s. Early attempts to preserve endangered species included efforts to maintain the heath hen prior to the Civil War, efforts to protect bison starting in the 1870s, and establishment of the Aransas National Wildlife Refuge in 1937 to protect the whooping crane.

At the end of the nineteenth century, the states' role as the nation's wildlife managers seemed firmly in place. A U.S. Supreme Court decision in 1896, *Geer v. Connecticut*,[2] squarely positioned the individual states to regulate the "common property" of game, including setting the conditions for the taking of game. However, the federal government's powers, granted by the Constitution, would gradually redistribute some of the states' management authority.

FEDERAL INTERVENTION

Until the early twentieth century, there were few conflicts between federal and state wildlife managers. Following the birth of the environmental movement, however, and the work of Roosevelt, Muir, Audubon, and others, the federal government began to exercise its constitutional powers (Bean, 1983). The first federal wildlife management legislation was the Lacey Act of 1900.[3] Relying on the commerce power in the Constitution, Congress prohibited interstate transport of wild animals killed in violation of state law. The Lacey Act was a cautious first step by the federal government. It was designed more to buttress enforcement of state regulation, however, than to assert federal management authority. Meanwhile, the decimation of waterfowl populations continued to confound state management. Hunting restrictions enacted by one state simply left more birds for hunters in other states. Such conflicts led to the first attempt to bring all migratory game birds under federal control. The effort started in 1904 and culminated in passage of the Migratory Bird Treaty Act in 1918.[4]

Since passage of the Migratory Bird Treaty Act, a range of significant federal laws affecting state management have been enacted—notably the Bald Eagle Protection Act in 1940,[5] the Wild Free-Roaming Horses and Burros Act in 1971,[6] the Marine Mammal Protection Act of 1972,[7] and the Endangered Species

Act of 1973.[8] Many of these wildlife laws have been the subject of states' rights debates. In the Marine Mammal Protection Act of 1972, Congress first demonstrated its willingness to preempt state authority over the taking of species. This paved the way for the inclusion of restrictions on taking in the Endangered Species Act of 1973. (Previous versions of the ESA did not include taking restrictions.) The Endangered Species Act of 1973 took away state control of taking for species listed under the act, setting in place an uneasy relationship between state and federal wildlife agencies. Like unwelcome in-laws, the federal government stepped in to work with the states on these key species. While working with the state agencies, the federal government retains control over the purse strings and the ultimate control—taking.

SECTION 6 COOPERATIVE AGREEMENTS

With the passage of the ESA, the federal government took primary responsibility for protecting endangered species. As part of the act, however, Congress also provided the opportunity for states to play a role in managing their resident threatened and endangered species and to receive federal financial aid for carrying out that role.[9] Section 6 funding can be viewed as the salve used to make state/federal cooperation under the ESA more workable. This funding produces greater protection for species and greater information on what is needed to protect them in the long run.

The conference report accompanying the adoption of the ESA in 1973 recognized the importance of developing state/federal cooperation:

> The successful development of an endangered species program will ultimately depend upon a good working arrangement between the federal agencies, which have broad policy perspective and authority, and the state agencies, which have the physical facilities and the personnel to see that state and federal endangered species policies are properly executed. The grant program authorized by this legislation is essential to an adequate program.... The conferees wish to make it clear that the grant authority must be exercised if the high purposes of this legislation are to be met.[10]

Under Section 6 of the act, states are encouraged to enter into cooperative agreements with the federal government. In fact, the interior secretary *must* enter into a cooperative agreement with a state unless he or she determines that the state's proposed program is not in accordance with the ESA's criteria.[11] With a cooperative agreement in place, a state is then eligible to receive funds. The federal government will pay up to 75 percent of program costs for agreements involving one state and up to 90 percent of costs when two or more states work cooperatively. The act separates plant agreements from animal agreements. Forty-eight states currently have a signed cooperative agreement of some type for endangered animal species and are thus eligible to receive Section 6 funds. Of these, thirty-six states also have an agreement signed for plant species. Only Alabama and Louisiana remain without any kind of cooperative agreement. (See Figure 1.)

Section 6 cooperative agreements have been a part of many notable ESA successes. Cooperative agreements have helped recovery efforts for the peregrine falcon in several states. They have provided funding for research on the critically endangered black-footed ferret. They have resulted in the reintroduction of the red wolf in North Carolina. And they are currently being used to implement hundreds of other projects to protect habitat and assist in the recovery of threatened and endangered species.

Despite the handful of success stories, the Section 6 cooperative agreement program is perennially plagued by inadequate and inconsistent funding. Table 1 shows recent funding levels for Section 6 grants to state programs. Yet these figures belie the true story. Viewing the ESA as an impediment to economic progress, the Reagan administration used the budget process to undermine it (Drabelle, 1985). Under President Reagan, the administration frequently (five out of seven budget years) requested no funding at all for Section 6 grants to states under the ESA. Each year, Congress restored funding for Section 6 agreements, but the amounts were minimal and uncertain, severely weakening the program as a whole. Because the recovery of endangered species is often a long and arduous process, endangered species research requires consistently funded multiyear efforts. Biologists need to develop long-term plans to research and implement recovery. As a result of the persistent funding

FIGURE 1

CURRENT STATES WITH COOPERATIVE AGREEMENTS

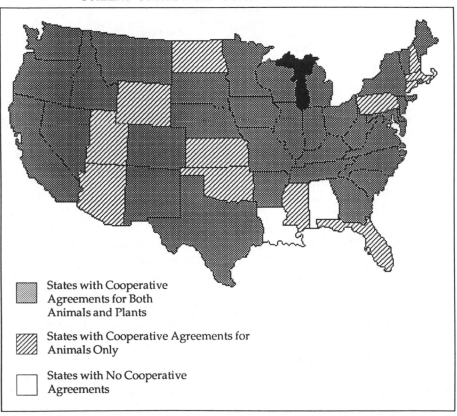

States with Cooperative
Agreements for Both
Animals and Plants

States with Cooperative Agreements for
Animals Only

States with No Cooperative
Agreements

uncertainties, however, states have reduced their long-term projects because they fear they will be prematurely terminated by unforeseen budget changes. Some states, notably Ohio, have eliminated their requests for Section 6 grants altogether.[12]

Recognizing Section 6 funding problems, Congress amended the ESA during reauthorization in 1988. The 1988 amendments established a cooperative fund from which grants are appropriated for the Section 6 program. Monies are allocated from general revenues into the fund in an amount equal to 5 percent of the total of Pittman-Robertson and Wallop-Breaux federal aid accounts for state programs.[13] The endangered species fund does not take any money from either of these federal aid pro-

TABLE 1

**RECENT FUNDING FOR SECTION 6
GRANTS TO STATES**

Fiscal Year	Dollars (thousands)
1982	0
1983	2,000
1984	2,000
1985	3,920
1986	4,204
1987	4,300
1988	4,300
1989	5,000
1990	5,676

grams; rather, it uses these programs as an index. The fund remains subject to annual appropriations, however, so the problem of reliable long-term funding remains a pivotal issue.

The Senate report accompanying these amendments emphasized the inadequacy of funding for cooperative state/federal programs. Funds appropriated in 1987 were roughly the same as in 1977, but cooperative agreements with the states increased fourfold in the same period. In 1977, therefore, Section 6 funding could provide approximately $200,000 for each of the twenty-one cooperative agreements in effect. In 1987, funding could provide only about $57,000 per agreement for the seventy-six agreements in effect.[14]

THE TAKING PROVISION

Much of the ESA's strength lies in its almost complete restriction on taking an endangered species.[15] Under the act, the meaning of the term "take" includes "harass, harm, pursue, hunt, shoot, wound, kill, trap, capture, or collect, or to attempt to engage in any such conduct."[16] This is perhaps the strongest and most far-reaching provision of the ESA. The word "harm" in the definition of taking has been further defined by regulation to mean "an act which actually kills or injures wildlife. Such act may include significant habitat modification or degradation where it actually kills or injures wildlife by significantly impairing essential behavioral patterns, including breeding, feeding, or sheltering."[17] While federal agencies are prevented from jeopardizing a species under Section 7 of the act, state agencies are not covered directly by this provision. The ESA does prevent state agencies from taking a listed species, however, either directly or indirectly.

In *Palila v. Hawaii Department of Land and Natural Resources*,[18] the court agreed that the state's management program for feral sheep and goats resulted in the destruction of nesting habitat for the palila, an endangered Hawaiian bird, and thus violated the taking provision of the ESA (Bean, 1983). In addition to testing the interpretation of habitat modification as a form of taking, the palila case further defined the constitutional basis for federal intervention in state management for endangered species. The state contested the applicability of the ESA to a bird that existed solely on state lands within one state. The court upheld the provision, however, relying on both the treaty and commerce powers held by the federal government. In addition, the court suggested that the importance of preserving a national resource such as an endangered species may even "be of such magnitude as to rise to the level of a federal property interest."[19]

For species listed under the ESA, management often requires only the identification and protection of habitat. But management of some species may require "taking," and thus the ESA's strong federal intervention into taking becomes a focus of controversy. Species that may cause injury to life or property, even while existing in endangered numbers, pose this paradox of taking before recovery and delisting. State and federal wildlife managers are struggling to reconcile this quandary, and again the most visible signs are in Yellowstone National Park.

GRAY WOLVES AND GRIZZLY BEARS

Large predators have become a focal point for conflicts between state and federal wildlife managers. In Yellowstone National Park, efforts to reintroduce the gray wolf and protect grizzlies have attracted considerable public attention. Large predators are unique in threatened and endangered species management because individuals within a population may affect humans by killing livestock or threatening hikers. These "problem individuals" are often taken from the population, even while the population as a whole requires protection. Taking these animals often means killing them, for translocation of problem individuals is considered largely ineffective for wolves (Fritts, 1982) and only moderately effective for grizzly bears (Harting, 1985). The very suggestion of a program that may involve regular taking of animals being controlled by the federal government has reignited deep-seated federalist tensions. Because states traditionally have controlled taking through sport hunting and trapping seasons, they often view the federal endangered species program as a usurpation of their authority.

The issue of taking listed predators first came to prominence in the controversy over wolf trapping in Minnesota. Two court cases set precedents that were immediately applied to grizzly bear hunting seasons in Montana and to wolf recovery in the West (Ernst and Warner, 1989). A 1978 court decision, rendered in *Fund for Animals v. Andrus*,[20] determined that the U.S. Fish and Wildlife Service (FWS) was taking nondepredating wolves in violation of its own regulations. Later, environmentalists challenged FWS regulations issued in 1984 that allowed sport trapping of Minnesota wolves. In a landmark decision, *Sierra Club v. Clark*,[21] the court determined that the proposed sport trapping of threatened wolves was unlawful. In its ruling, the district court distinguished between protection for species listed as threatened (versus endangered) by citing the ESA provision that allows the taking of threatened species under circumstances of extraordinary population pressures.[22] In defending its actions, however, the FWS made no attempt to demonstrate that sport trapping regulations were needed to relieve population pressures.

The lack of further judicial definition of extraordinary population pressures fuels the controversy over wolf and grizzly bear management in the West. Many state wildlife managers believe the ruling in *Sierra Club v. Clark* unreasonably limits options for managing endangered and threatened predators (Schwinden, 1987). The Montana Department of Fish, Wildlife, and Parks (MDFWP) resisted the original listing of the grizzly bear and still retains a vestigial sport hunting season on the threatened bear. It does so, however, under threat of a potential ESA lawsuit. Fears that the *Sierra Club v. Clark* decision might raise questions about the legality of northern Montana's grizzly bear hunt helped fuel an unsuccessful attempt to overturn the 1985 decision. Meanwhile, the state of Wyoming has requested a change in regulations that would allow taking of nuisance grizzly bears by hunters.

Montana also wants to modify delisting criteria for the gray wolf. State officials support wolf recovery, including reintroduction into Yellowstone National Park, only if the wolf is delisted. Arguing that the reintroduction of wolves will result in a loss of livestock and big game, state biologists demand more options in controlling wolves before they will support the recovery plan. Assuming that recovery in northern Montana will occur much more rapidly than in Yellowstone, the MDFWP is concerned that the northern population will reach unacceptable levels before the wolf is delisted. MDFWP suggests that each of the three recovery areas be treated separately. Accordingly, the 1987 recovery plan for the wolf does call for downlisting to threatened status for a population of wolves that has met the recovery goal for a particular recovery area (U.S. Fish and Wildlife Service, 1987).

Because of conflicts over how wolves should be managed, Montana and Wyoming have excluded themselves from participating in the recovery process. (See Robbins, 1986; Zumbo, 1987.) As well, the Idaho legislature has forbidden the use of any state funds for wolf recovery efforts. Indeed, the wolf and grizzly bear taking controversy has spilled over into Congress. Although Congress as a whole has shown overwhelming support for the Endangered Species Act,[23] delegations from western states have made efforts to reshape the ESA and stall its implementation— particularly the wolf recovery plan. That the 1988 reauthorization was held up in the Senate for three years was due partly to concerns of western senators over the management of wolves and grizzly bears. In 1985, Senator Alan Simpson (R–WY) helped to block reauthorization of the ESA because of his concerns over the ruling in *Sierra Club v. Clark*. Reportedly Senator Simpson wanted an amendment that would allow the hunting of grizzly bears in Wyoming. He also wanted a provision that would permit ranchers to protect livestock from wolves that cross the boundaries of Yellowstone National Park (Stern, 1988). While no amendment concerning grizzly bear hunting was added to the act, the report by the Senate Environment and Public Works Committee that accompanied the 1988 reauthorization of the ESA declared the committee's impression that the regulated harvest of grizzly bears in Montana is consistent with the act.[24] Similarly, in the House, Representative Ron Marlenee (R–MT) unsuccessfully sought to pass an amendment to the act during floor consideration in December 1987 that would have removed the gray wolf from the list of endangered species.

While not changing the Endangered Species Act, the three members of Wyoming's congressional delegation have been successful in frustrating its implementation in general and the gray wolf's recovery in particular. Through political pressure on the Reagan administration, they postponed the environmental impact statement (EIS) for the evaluation of reintroduction of wolves into Yellowstone (see Fischer, 1987). Frank Dunkle, director of the U.S. Fish and Wildlife Service from 1986 to 1989, equivocated repeatedly on wolf reintroduction. Western newspapers reported him supporting recovery at one meeting and then opposing it weeks later at another meeting (Williams, 1988). Regardless of Dunkle's contradictory rhetoric, wolf recovery did not progress past the release of the recovery plan in 1987.

In response, Representative Wayne Owens (D–UT) began to push the administration forward on wolf reintroduction. He first sponsored a bill in 1987 that required the restoration of wolves to Yellowstone within three years. In 1989, Owens modified his bill to mandate an EIS on wolf reintroduction. His goal has been to bring the issue to a debate in Congress (Owens, 1988). Thus far, he has been successful in obtaining a hearing on the bill. In July 1989, wolf advocates and opponents gathered to testify on the proposed legislation. Significantly, at this hearing, Wyoming Senator Alan Simpson indicated, while opposing Owens' bill, a willingness to discuss a solution. As a counter to Owens' bill, Senator James McClure (R–ID) offered legislation in 1990 that would reintroduce wolves into recovery areas in Idaho and Wyoming—under the condition that wolves would be delisted outside these areas.

McClure's proposal strikes at the center of the wolf recovery issue. By delisting the wolf outside the recovery areas, the boundary line between state and federal control of taking is clearly drawn—literally on a map. The federal government would retain full control of taking within the recovery area where the wolf would remain listed, and the state would have control outside recovery areas. One question still to be answered is whether state wildlife management agencies would be able to wrest control over wolf recovery away from agricultural interests who may simply want a strict trapping program for any wolf that strays from the recovery area.

Meanwhile, the presidential administration has changed hands. In 1989, President Bush appointed John Turner, a former Wyoming state senator who lives in the Yellowstone area, as the new director of the Fish and Wildlife Service. Turner, a seasoned politician with a strong environmental record, is clearly familiar with the wolf recovery issue. It is still too early, however, to predict whether he will be able to break the stalemate.

ISSUES AND PROSPECTS

As natural resources have been depleted, concern for wildlife has caused the federal government to become increasingly involved in protection of wildlife resources—especially for species of national concern such as those in danger of extinction. While, in general, states have benefited from the additional support for endangered species protection, they have not welcomed the federal preemption of their traditional control of the wildlife resource.

Implementation of the Endangered Species Act has been caught in a sometimes uncomfortable alliance between states as managers of wildlife and the federal government as protector of endangered species. That alliance is most strained when protection of endangered species encroaches on the state's authority to control taking. Yet the alliance is necessary. Assuming that state agencies have ignored the plight of the grizzly bear would be just as naive as assuming they do not need the federal support provided by the ESA to protect the bear's habitat. Likewise, assuming that federal agencies have ignored the needs of state agencies would be just as naive as assuming they do not need the state agencies' support to recover endangered species such as the wolf and grizzly bear.

The ESA meets reality at the juncture of state and federal management. The lofty goals of the act run headlong into the reality of wildlife management—no species lives in isolation. Species such as the gray wolf and grizzly bear require managing a large area of land in which contingencies such as livestock depredation are likely to occur. Clearly an argument can be made that introducing wolves into Yellowstone will benefit both the wolf and the Yellowstone ecosystem. But it is also true that wolves may stray out of this ecosystem to dine on livestock. Debate over the actual extent of this threat will continue indefinitely. The question is: How will managers respond if wolves do need to be controlled? This is an issue that state managers are reluctant to entrust to federal wildlife managers.

Senator McClure's proposal to reintroduce the wolf, but to delist it outside a defined recovery area, raises interesting questions. The proposal draws clear lines between state wildlife managers, who would have authority outside recovery areas, and federal agencies with authority inside these areas. If the recovery area is of sufficient size and encompasses appropriate habitat, the wolf could be reestablished and state agencies would have free reign on their side of the fence. But the proposal assumes that state and federal managers are at cross-purposes—federal agencies interested in protecting endangered species and state agencies concerned with game species and livestock. This is clearly an oversimplification. While it may solve the problem for livestock owners, it is not a complete solution for big game management. Since big game migrate freely, management of big game outside the fence will depend on management of big game and wolves inside the fence. Likewise, wolf control efforts outside will depend on the wolf population inside. While McClure's proposal may ultimately be workable, it does not change the need for state and federal wildlife managers to cooperate.

Notes

The views of the author do not necessarily reflect those of the National Wildlife Federation.

1. 41 U.S. (16 Pet.) 367 (1842).

2. 161 U.S. 519 (1896).

Federalism and the Act 111

3. Ch. 553, 31 Stat. 187 (May 24, 1900) (current version at 16 U.S.C. §§701, 1540, 3371–78 and 18 U.S.C. §42).

4. Ch. 128, 40 Stat. 755 (July 3, 1918) (current version at 16 U.S.C. §§703–711).

5. 16 U.S.C. §§668–668d (1976 & Supp. V 1981).

6. 16 U.S.C. §§1331–1340 (1976 & Supp. V 1981).

7. 16 U.S.C. §§1361–1407 (1976 & Supp. V 1981).

8. 16 U.S.C. §§1531–1543 (1976 & Supp. V 1981).

9. 16 U.S.C. §1635 (1986).

10. Conf. Rep. No. 740, 93rd Cong., 1st Sess. (1973).

11. 16 U.S.C. §1635(c)(1) (1986).

12. Sen. Rep. No. 240, 100th Cong., 2nd Sess. 5 (1987).

13. Pittman-Robertson and Wallop-Breaux accounts refer to the Federal Aid in Wildlife Restoration Act (enacted in 1937) and the Federal Aid in Fish Restoration Act (enacted in 1950 and amended by the Wallop-Breaux amendments of 1984), respectively. Both acts establish a fund derived primarily from federal excise taxes on hunting and fishing equipment. The money in these funds is apportioned to the states based on geographic area and numbers of hunting and fishing licenses sold.

14. Sen. Rep. No. 240, 100th Cong., 2nd Sess. 5 (1987).

15. Section 10 authorizes exceptions to the prohibition on taking endangered species "for scientific purposes or to enhance the propagation or survival of the affected species"; 16 U.S.C. Sec. 1539 (a)(1)(A) (1986).

16. 16 U.S.C. Sec. 1532 (19) (1986).

17. 46 Fed. Reg. 54748, 54750 (Nov. 4, 1981).

18. 471 F. Supp. 985 (D. Hawaii 1979).

19. 471 F. Supp. at 995 n. 40.

20. 11 Env't Rep. Cas. (BNA) 2189 (D. Minn. 1978).

21. 577 F. Supp. 783 (D. Minn. 1984), aff'd, 755 F. 2d 608 (8th Cir. 1985).

22. Notably, species listed as threatened are not afforded direct statutory protection as are endangered species. Rather, under Section 4(d) of the ESA, the secretary of the interior is directed to "issue such regulations as he deems necessary and advisable to provide for the conservation of [threatened species]"; 16 U.S.C. §1532 (3)(1986). "Conservation" in the ESA means "to use and the use of all methods and procedures which are necessary to bring any en-

112   COMPONENTS OF THE FEDERAL ENDANGERED SPECIES PROGRAM

dangered species or threatened species to the point at which the
measures provided pursuant to this Act are no longer necessary."
Such methods "in the extraordinary case where population pressure within a given ecosystem cannot be otherwise relieved, may
include regulated taking"; 16 U.S.C. §1532 (3)(1986).

23. The ESA's most recent reauthorization in 1988 passed the House
by a vote of 399 to 16 and passed the Senate by a vote of 93 to 2.

24. Sen. Rep. No. 240, 100th Cong., 2nd Sess. 5 (1987).

References

Aderhold, M. 1987. "All for the Wolf." *Montana Outdoors* 18(5):15–22.

Anderson, Stanley H. 1985. *Managing our Wildlife Resources.* Colum-
bus, Ohio: Bell & Howell.

Bean, Michael. 1983. *The Evolution of National Wildlife Law.* New York:
Praeger.

Bean, Michael. 1986. "The Endangered Species Program." Pp. 347–371
in R. L. DiSilvestro (ed.), *Audubon Wildlife Report 1986.* New York:
National Audubon Society.

Drabelle, D. 1985. "The Endangered Species Program." Pp. 73–90 in R.
L. DiSilvestro (ed.), *Audubon Wildlife Report 1985.* New York: National Audubon Society.

Ernst, John P., and Susan H. Warner. 1989. "Wolf and Grizzly Bear
Management in the Western United States: Where Do We Go from
Here?" In *Proceedings IV: Issues and Technology in the Management of
Impacted Western Wildlife.* Boulder: Thorne Ecological Institute.

Fischer, H. 1987. "Deep Freeze for Wolf Recovery." *Defenders*
62(6):29–33.

Fitzgerald, J. 1988. "Withering Wildlife: Whither the Endangered Species Act?" *Endangered Species Update* 5(10):27–34.

Fritts, S. 1982. *Wolf Depredation on Livestock in Minnesota.* U.S. Fish
and Wildlife Service Research Publication 145. Washington, D.C.:
Government Printing Office.

Harting, A. L. 1985. "Relationships Between Activity Patterns and
Foraging Strategies of Yellowstone Grizzly Bears." Master's thesis,
Montana State University.

Owens, W. 1988. "Crying Wolf at Yellowstone." *National Parks*
62(3–4):16–17.

Robbins, J. 1986. "Wolves Across the Border." *Natural History*
95(5):6–15.

Schwinden, T. 1987. Letter to Galen Buterbaugh, Regional Director, U.S. Fish and Wildlife Service, Denver, Feb. 25, 1987.

Stern, A. 1988. *Congressional Quarterly Weekly Report*, July 30, 1988, 46(31):2113.

Trefethen, James B. 1975. *An American Crusade for Wildlife*. New York: Winchester Press.

U.S. Fish and Wildlife Service. 1987. *Northern Rocky Mountain Wolf Recovery*. Denver: U.S. Fish and Wildlife Service.

Williams, T. 1988. "Confusion About Wolves." *Gray's Sporting Journal* 13(3):132–156.

Zumbo, J. 1987. "Should We Cry Wolf?" *Outdoor Life*, Dec. 1987, pp. 50–100.

INTERNATIONAL IMPLEMENTATION: THE LONGEST ARM OF THE LAW?

by

MARK C. TREXLER AND LAURA H. KOSLOFF

NOTWITHSTANDING THE VARIETY of implementation problems outlined elsewhere in this volume, the 1973 Endangered Species Act (ESA) is the envy of wildlife conservationists around the world. Its habitat protection and consultation requirements go far beyond provisions found in most endangered species legislation. The problem of species extinctions, however, is much worse outside the United States. Thousands of species of fauna, tens of thousands of species of vascular plants, and huge numbers of invertebrate species are endangered or threatened worldwide. Six hundred million years of gradually increasing species diversity, particularly in the tropics, are coming to an end (Vermeij, 1986).

Drafters of endangered species legislation in the United States were aware early on of the geographic imbalance of the extinction problem. Determined at the very least that the United States should not be responsible for the extinction of species abroad, the 1969 Endangered Species Conservation Act put the United States in the business of promulgating import restrictions based on its own regulatory findings regarding a foreign species' status.[1] Unilateral trade restrictions were seen as insufficient and competitively disadvantageous to U.S. wildlife traders, however. The 1969 act therefore directed the Departments of State and Interior to host a ministerial meeting with the aim of arriving at a "binding international convention on the conservation of endangered species."[2]

The eventual outgrowth of this mandate, the Convention on International Trade in Endangered Species of Wild Fauna and Flora (CITES), has become the most important component of this country's international wildlife conservation efforts within the context of the ESA. Moreover, the Endangered Species Act today continues the listing of foreign species on the U.S. threatened and endangered species list, provides for the implementation of the Western Hemisphere Convention, authorizes U.S. federal agencies to assist foreign countries in improving their wildlife management, and imposes conditions on U.S. agency actions taken abroad. This essay reviews the provisions of the ESA relating to U.S. international responsibilities for wildlife conservation and evaluates their effectiveness.

INTERNATIONAL APPLICATION OF SECTION 7 CONSULTATIONS

Section 7 of the ESA requires federal agencies to consult with the Fish and Wildlife Service (FWS) to ensure that any action they undertake is "not likely to jeopardize the continued existence" of a listed threatened or endangered species.[3] International application of this provision could serve as a strong symbolic statement of this country's commitment to international wildlife conservation. Indeed, the Department of the Interior's original regulations implementing this provision required consultation for any federal agency action, including those taken abroad.[4] The regulation's international application suffered from a severe case of benign neglect, however. Finally, in 1986, new Section 7 regulations deleted federal actions occurring outside the United States from the consultation requirement altogether.

Defenders of Wildlife challenged the deletion in federal district court. The court, siding with the environmental group, recently held that the consultation provisions of Section 7 clearly apply to federal actions undertaken in foreign countries. The decision was recently affirmed in the Eighth Circuit.[5] In the interim, the FWS stands by its 1986 regulations.

It is not clear, however, that a favorable judicial decision would have a significant practical effect. Few consultations involving federal actions abroad were initiated between 1978 and 1986 even when Interior's policy required them. The number of species able to force such a consultation remains rather small.[6] Moreover, FWS staff is spread thinly in the face of dramatic increases in informal domestic consultations,[7] and there are significant practical and political obstacles facing effective implementation of the requirement in foreign countries. Nevertheless, it remains an important symbolic provision of the act and should be reinstated.

INTERNATIONAL ASSISTANCE THROUGH SECTIONS 8 AND 8A(E)

Although requiring U.S. federal agencies to consult on actions taken abroad might demonstrate this country's commitment to international wildlife conservation, two other provisions of the ESA are potentially more practical. The first of these provisions, Section 8, allows the FWS to provide assistance in support of foreign fish, wildlife, and plant conservation programs. Section 8 generally allows for training and educational assistance and has been used for such diverse activities as the production of television programs and the establishment of wildlife clubs. A subsection of Section 8 authorizes the Interior Department to use foreign currencies owned by the United States to aid species on the U.S. endangered and threatened list.

Section 8 has been applied narrowly, however, particularly during the Reagan years. Although funds have been committed to various educational and training programs, the dollar amounts have been small. The total funding for environmental education programs under Section 8 depends on FWS's discretionary funds; for fiscal year (FY) 1989, funding totaled only $20,500. That amount has not varied much in recent years. Some $510,000 has accumulated in the excess foreign currency program and is committed to projects in Egypt, Pakistan, and India for FY 1989. Over a ten-year period, available funds in the program have totaled approximately $2 million in these three countries—an amount that precludes effective development of the wildlife management agencies in countries that need assistance.

The second provision, Section 8A(e), provides for implementation of the 1940 Western Hemisphere Convention. Prior to 1982, this section provided only for U.S. representation in matters involving the convention. Since 1982, Section 8A(e) has directed the FWS to provide technical and training assistance to foreign governments and to develop measures to protect migratory bird and plant species covered by the treaty. Funding for eleven projects under Section 8A(e) for FY 1989 amounts to approximately $500,000, an increase from the initial congressional appropriation of $150,000. These projects include such activities as training wildlife biologists, establishing a resource center for graduate students, operating wildlife management workshops, and providing assistance for environmental education workshops. The treaty as a whole, however, is perceived by many observers as a "sleeping treaty" that in practice has accomplished little (Lyster, 1985).

THE CONVENTION ON INTERNATIONAL TRADE IN ENDANGERED SPECIES

By any measure, the international wildlife trade is an extraordinarily complex economic activity. The trade supports a multibillion-dollar industry involving dozens of countries, thousands of species, hundreds of thousands of shipments, and millions of participants.[8] Through this trade, fewer than 10 million reptile skins become 30 million manufactured products each year. Some 80,000 elephant tusks become 28 million manufactured ivory products. The United States alone imports more than 200 million manufactured products every year, in addition to hundreds of millions of live specimens, mostly captive-bred tropical fish.

Few data exist concerning the historical scope, development, and composition of the trade. Since World War II, the technical potential for detrimental trade has grown significantly with the development of new exploitative technologies, the incursion of human populations into remote areas, and improved global communications. Today it is possible not only to exploit truly rare and formerly inaccessible species but to rapidly and even inadvertently overexploit common species. Moreover, other perils facing species have worsened in recent decades. As a result, threats currently posed by habitat loss and degradation, introduction of exotic species, and domestic overexploitation continue to dwarf the detrimental impact of the international wildlife trade for most species. But because of the wildlife trade's perceived impact on species of aesthetic or special interest, many groups are attracted to the concept of controlling it. Combined with its suggested susceptibility to international action, the role of trade controls in international conservation efforts has come to outweigh by far the trade's role in threatening species.

An Overview of CITES

CITES' ambitious goal is to regulate the complex wildlife trade by controlling species-specific trade levels on the basis of biological criteria. The treaty establishes three appendices. Appendix I is to include all species "threatened with extinction which are or may be affected by trade." CITES generally prohibits commercial trade in these species. Appendix II is to include all species that "although not necessarily now threatened with extinction may become so unless trade in specimens of such species is subject to strict regulation." "Look-alike" species (species that could be subject to increased trade pressure due to their resemblance to threatened species already listed) are included in Appendix II. While the intent of Appendix I is straightforward—to prevent wildlife trade from causing species extinctions through the enforcement of trade bans—the goal underlying Appendix II reflects a much more sophisticated task. That task is to allow trade in any listed species as long as the transaction will not prove detrimental to the species' ability to play its natural biological role throughout its range. Appendix III lists species at the request of a country in which the species occurs and in which the species is protected; its goal is to help countries enforce domestic protection.

CITES' regulatory framework consists of specific permitting requirements associated with each appendix. By requiring a finding that a particular transaction will not cause any detriment to the survival of the species in the wild (the "no-detriment" finding) prior to the issuance of any trade permits for Appendix I or II specimens, CITES attempts to link treaty implementation directly to the conservation of species in the wild. By requiring that countries only allow the import and export of Appendix I and II shipments when a paperwork trail back to the original no-detriment finding legitimizes the shipment, CITES attempts to keep illegal trade from bypassing the no-detriment requirement. To implement these basic provisions, CITES creates an international secretariat, requires the parties to meet biennially in a policy-setting Conference of the Parties (COPs), and charges the parties with specific implementation responsibilities. CITES' regulatory success is premised on achievement of a universal level of implementation of its provisions. "Weak-link" parties, for example, can undercut the implementation efforts of other parties by inappropriately issuing "legitimate" paperwork.

U.S. CITES Responsibilities

CITES requires each party to establish offices to confirm that regulated specimens were taken or imported legally, that transactions will not be detrimental to the species, and that specimens are properly transported. Border control agencies in turn are to verify, prior to any export, import, or reexport of CITES specimens, that the shipment is accompanied by valid paperwork. Shipments in violation of CITES are to be confiscated and the traders penalized. Each party is to maintain records of trade in CITES specimens and submit detailed annual trade reports to the secretariat.

CITES is implemented in the United States through the Endangered Species Act. The two sections of the ESA that implement CITES, Section 8A and Section 9(c), have resulted in truly significant resource commitments by the U.S. government relative to the ESA's other international components. Section 8A directs the Interior Department to carry out the United States' CITES obligations, while Section 9(c) makes CITES violations punishable under the ESA. Millions of tax dollars are spent each year on activities relating to CITES, ranging from support of the secretariat to funding of the FWS wildlife inspector program to punishment of CITES violations.[9] As a result, the United States is considered the best implementer of CITES among the 102 parties.

Notwithstanding its role in fulfilling the congressional mandate to organize a "binding international convention on the conservation of endangered species," one must be realistic about what CITES sets out to do. As a trade control instrument, CITES' ability to benefit species is limited. Relatively few species are significantly threatened by international wildlife trade alone. Controlling international trade in a species whose habitat will soon disappear or that will succumb to domestic overexploitation is a hollow victory at best.

CITES sets out only to prevent detrimental international trade in endangered species. It does not provide for habitat protection, control the introduction of exotic species, or restrict the overexploitation of domestic species. It does not require species recovery plans or any sort of consultation process. It does not attempt to control the demand for the products of endangered species.

CITES did cap a long line of international legal instruments aimed at wildlife conservation, however, most of which contained trade control provisions. Nongovernmental organizations (NGOs) jumped at the chance to draft a treaty focusing exclusively on wildlife trade controls, continuing to believe that controlling the wildlife trade would be relatively easy to achieve. Governments were also enthusiastic about this approach. After all, agreeing to control the international wildlife trade poses few national sovereignty concerns, as would habitat conservation and even domestic exploitation restrictions, and involves few difficult political or economic choices for national decision makers. These facts go a long way toward explaining how the goal of international endangered species conservation expressed in the 1969 act ultimately translated into a narrow trade control treaty.

THE EXPERIENCE THUS FAR

The impact of CITES thus far is disappointing even if one limits one's goals to CITES' own narrow scope (Trexler, 1990). The behavior of CITES parties suggests the pursuit of rhetorical aims having little to do with substantive species conservation. Moreover, CITES' divergence from a course that might better achieve species conservation aims is increasing rather than diminishing.

Even if it were implemented according to its terms, CITES' ability to improve the biological status of species in the wild is questionable. One inherent problem is that CITES is built on a set of ideal assumptions that seem rational but are unconvincing when looked at in detail. In addition to assuming that trade control procedures can pinpoint illegal and biologically detrimental trade, CITES assumes that successfully implemented trade controls will result in fewer specimens of endangered species being taken from the wild. It also assumes that successfully implemented trade controls will translate into reduced demand, rather than an uncontrollable black market for wildlife products. Both assumptions are riddled with exceptions.

Even beyond CITES' permit structure, the development and implementation of CITES would have to be carefully coordinated if it were to stand a significant chance of actually benefiting listed species. Substantial information must be collected to justify the appendices and make daily permit decisions; competent bureaucracies must be established to administer the trade controls; careful targeting of limited enforcement expertise and resources is needed to match enforcement efforts with those species that might actually benefit from trade controls; and strict punishment of violations is needed to deter illegal trade.

Practical Difficulties

Regulatory theory and past experience with international regulatory instruments suggest that controlling international wildlife trade for conservation purposes will be difficult, even for the species that CITES could benefit. A skin that is the product of illegal overexploitation looks no different when crossing international borders than a skin produced by a rational (and legal) management regime. It is extremely difficult to distinguish biologically detrimental trade from nondetrimental wildlife trade and from the huge international flow of goods generally. If specimens are being smuggled, they will likely never be noticed at all, as the following scenario illustrates: "Its unmarred skin was drying on a rack within hours. Within a week, it was smuggled by plane across the border [from Brazil] to Paraguay where, since jaguars are protected by Paraguayan law, it was packed into a crate labeled 'Café do Brazil.' The sheer volume of cargo to be checked in customs prevented anyone from checking the crate's contents. It was shipped to West Germany undetected" (World Wildlife Fund, 1985).

Even if one discounts the problems posed by wildlife smuggling, the difficulty of detecting illegitimate trade going through normal commercial channels cannot be underestimated. Assuming that a border control agent has time to inspect a newly arrived shipment of caiman products, for example, he must verify that the permit and commercial invoice match the contents of the shipment. The agent must be able to differentiate between the skins and products of closely related species, since the import of one subspecies might be legal while that of another is not. He must confirm the legitimacy of the listed country of export, the permit's apparent authenticity, and the authorizing signatures. He must assure that there has been no recent change in domestic or foreign laws affecting trade in the species. He may have to evaluate the shipment's compliance with complex criteria regulating the permitting of the shipment's contents, such as captive-bred specimens. Or he may have to verify the shipment's compliance with quotas established for trade in the species and ensure that each specimen is appropriately marked if so required.

This is not what actually happens. With over one hundred parties, the variation in implementing procedures is immense. No-detriment findings in principle are required whenever any specimen of more than 150,000 populations, subspecies, and species enters into trade. Yet little of the biological information required by the CITES regulatory process has been collected, and performing no-detriment findings down to the population level in the tropics remains virtually impossible. Although most parties have implementing agencies, many exist largely on paper. CITES responsibilities are often divided among more than one agency, resulting in bureaucratic conflict and international confusion. Most national agencies charged with CITES implementation are grossly underfunded and understaffed. As a result, CITES parties have generally come to accept that issuance of an export permit does not mean a legitimate no-detriment finding has been made. Yet the entire point of the regulatory chain established by CITES is premised upon this assumption.

In 1973, the scientific data needed to specify CITES' appendices did not exist; today, these data still do not exist. The wildlife trade is constantly changing in response to such variables as species availability, consumer fashions, public health policy, and national law. Customs services are outward rather than inward-looking and are predominantly charged with drug interdiction and revenue collection rather than wildlife conservation. Moreover, the countries given the greatest and most expensive responsibilities under CITES are those least likely to be able to pay for them—namely the tropical wildlife-exporting countries.

Growth of CITES Appendices

The inherent complexity of the international wildlife trade is not the only difficulty CITES has faced. The parties have sunk into a legalistic morass of treaty interpretation in defining the scope of the appendices. In some cases, they even fail to agree on their purpose. Rather than guiding and focusing national implementation efforts, the parties have diluted their resources and expertise by constantly increasing the scope of CITES' appendices. Completely contrary to its drafters' intent, CITES "protection" now extends to at least 30,000 species of plants and animals. One expert estimates that only 20 percent of Appendix I species may conform to the treaty's listing criteria; the proportion is far smaller for Appendix II (Jorgen Thompsen, TRAFFIC (U.S.A.), pers. com.). The great majority of listed species thus are not threatened by the wildlife trade but nevertheless consume limited administrative and enforcement resources. Accurate species-specific identification is becoming increasingly unlikely.

At the Conference of the Parties (COP), the parties have acted as if information availability were perfect and implementing resources unlimited. The appendices have been allowed to become increasingly complex, making effective implementation of CITES' paperwork and inspection provisions more difficult.

Several layers of criteria supplementing the treaty text now regulate many portions of the trade (such as captive-bred and ranched specimens). Ironically, each successive layer of regulatory complexity is usually inspired by the failure of the simpler procedures that preceded it. These failures in turn often resulted from the basic complexity of the task, as well as from shortages of information and resources that the parties have never tried seriously to correct.

CITES parties have come to recognize that the original enthusiastic listing of species on the treaty's appendices could have significant economic implications. But the response has been less than effective. The wish of African countries to trade commercially in species of crocodiles listed on Appendix I, for example, has proved detrimental to the development of the treaty. Political pressure has allowed populations of several species to be moved from Appendix I to Appendix II without scientific justification, resulting in a complex enforcement mosaic. Implementation of Appendix I trade bans, which should be conceptually straightforward even if difficult to implement, has become more and more complicated. New ground in appendix complexity was broken at the 1987 conference when, in defiance of CITES' provisions, the chemically marked wool of certain vicuña populations was downlisted to Appendix II, while all other parts and derivatives as well as the wool from all other populations were left on Appendix I. There is no reason to expect that CITES parties can apply such complex criteria to their trade control efforts.

Inadequate Enforcement

CITES has become almost impossible to enforce. There is little bureaucratic incentive for species-specific no-detriment findings or paperwork verification to be pursued when an entire family is listed on the same appendix. Hence enforcement efforts are mostly directed toward species of little conservation concern. One can also expect enforcement to focus on the most easily detectable violations, rather than the truly detrimental ones. As the universe of regulated transactions grows, this problem only gets worse.

The severity of penalties assessed through CITES enforcement activities and their deterrent value are difficult to estimate. The secretariat consistently complains about the inadequacy of penalties utilized by the parties, but no systematic data have been assembled. It is generally believed that wildlife smuggling or other illegal activities relating to endangered species are low-risk endeavors, reportedly leading some drug traffickers to switch jobs.

There is reason to believe, too, that enforcement of CITES within the United States is not serving a significant deterrent function (see Kosloff, 1990). In the vast majority of cases, the only penalty imposed on a violator is confiscation of merchandise. Moreover, it seems that enforcement efforts are not focusing on those portions of the trade potentially most detrimental to CITES species. Instead, most reported violations involve tourist souvenirs or other personal, rather than commercial, transactions.

Ambiguity of Objectives

When the concept that eventually became CITES was first discussed in 1961, different interest groups had varying perceptions of what the proposed treaty would accomplish. Perhaps not surprisingly, this ambiguity in the objectives being pursued has carried over into CITES implementation. Governmental as well as nongovernmental interest groups have used the treaty to pursue policy objectives other than the original aim of preventing nondetrimental trade. These diverse objectives include completely shutting down the trade, improving transport conditions for live specimens, ensuring the freedom of legal trade, monitoring the wildlife trade, and conserving biological diversity. These goals may be legitimate aims for different types of international trade controls. But as Australia noted at the third Conference of the Parties in 1981, different objectives can suggest varied and sometimes contradictory approaches to the interpretation and implementation of CITES' provisions. The unwillingness of the parties to come to grips with this fact has seriously impaired CITES' performance.

CITES' Impacts

No one has been able to show actual biological benefit to any Appendix I species resulting from CITES' trade controls. This does not mean that trade pathways and mechanisms have not been affected, nor that trade volumes have not declined for some species. Overall, however, CITES parties have proved unwilling or unable to control trade in listed species. CITES appears unable to control the export, and often the import, of any wildlife product valuable enough to justify trading or smuggling in small quantities (such as the products of rhinos, sea turtles, elephants, and leopards). Controlling such trade presents the same obstacles as controlling the international narcotics trade. Moreover, the relative priority of these trade controls and the resources dedicated to the task are minuscule. Little is known about the magnitude of the smuggling that does occur, but enforcement officials acknowledge that it is widespread. When someone takes the time to look, elaborate smuggling systems can be found almost anywhere (Nichol, 1987).

There is particularly little reason to expect that CITES as it has been implemented would have much effect on the biological status of Appendix II species. Few parties have the interest or ability to assess accurately the detrimental nature of individual transactions. Some countries have even interpreted the purpose of Appendix II listings to be monitoring rather than controlling the trade in these species.

Trade in Appendix II species has increased dramatically since CITES came into effect. The most famous Appendix II species over the last decade has been the African elephant; its status illustrates the practical failings of CITES' paperwork controls. As concern over its declining numbers has grown, the African elephant has received more attention at the policy-setting Conferences of the Parties than any other individual CITES species. The parties have passed numerous resolutions establishing guidelines for controlling the ivory trade. In 1985, the parties established a complex quota system intended to bring this trade under control. Yet just four years later, in spite of the earlier quota system, CITES parties recognized the still deteriorating status of the elephant and voted to place it on Appendix I, thus banning commercial trade in ivory. Elephants are just one of many species that have moved from Appendix II to Appendix I. CITES' ambitious goal of balancing conservation with sustainable economic exploitation simply has not been achieved.

Unable and unwilling to evaluate CITES' true biological and conservation impacts, the parties have increasingly focused on technical legalities. Yet it is unclear why conservationists should be concerned about violations of national trade bans or even of CITES itself if a technical violation has no negative implications for a species in the wild. The number of bans and other restrictions imposed at the national level since CITES came into force is daunting. Rather than strengthening national implementation, the parties have constructed an international regulatory house of cards that threatens to collapse of its own weight. Trade bans and CITES' increasing complexity impede, rather than improve, effective control of the trade. Moreover, the legality of a specific transaction has become exceedingly difficult to judge as CITES and national laws have become more complex, and the biological implications of these transactions are all but ignored.

LESSONS FROM CITES

CITES has generated a tremendous amount of regulatory activity over the past fourteen years and has created a large implementing bureaucracy. It has been a tremendous political success; perhaps this alone advances the interests of wildlife conservation. Elsewhere in this book, Michael Bean concludes that although the ESA has not yet achieved a great deal domestically, it has established a solid framework for future accomplishments. The same cannot be said for international implementation of the Endangered Species Act through CITES. Nevertheless, given the implementation problems caused by CITES' assumptions and the confounding complexity of international wildlife trade, it would be unfair to consider CITES a conservation failure if the parties were showing signs of learning from its problems. Instead all the same problems—and even new ones of the parties' own making—continue to surface.

Many observers, however, persist in characterizing CITES as

a substantive as well as a political success. But even if CITES is perceived as a success compared to prior multilateral wildlife conservation instruments,[10] one should be careful about extrapolating from CITES to other issues. The wildlife trade is politically and economically inexpensive and poses a fundamentally different regulatory and conservation problem than does habitat conservation. There is no assurance that any of CITES' attributes, even if successful, would contribute to the success of other instruments addressing other problems.

The prognosis for the future of CITES, therefore, is bleak. Given the current trends in deforestation, continued population growth, general environmental distress, budgetary limitations, and the CITES developments described earlier, it is hard to see how the Endangered Species Act through CITES can play much of a role in slowing extinction rates abroad. Incremental progress in CITES' geographic coverage, implementing legislation, and identification techniques should not be dismissed. Unfortunately, such progress will continue to be overwhelmed by weaknesses in the regulatory structure, growth of the wildlife trade, changing international trade patterns including the lowering of customs barriers, and implementation difficulties of the parties' own making.

CITES focuses on the international wildlife trade because it is believed that the domestic wildlife management practices of sovereign states are not open for international discussion. But whether its drafters intended it or not, CITES offers a means through which these national sovereignty barriers can be bypassed.[11] Properly staffed and supported, CITES implementing agencies in exporting countries could assume many wildlife management and conservation functions. By assisting with the no-detriment findings mandated by CITES, the international community can conduct biological and environmental research in party states unable or unmotivated to undertake such efforts on their own. The responsibility of parties to enforce CITES and punish violations opens the way for external assistance in redrafting and improving existing domestic wildlife legislation. It also provides avenues for the exchange of technical and policy advice on implementing such legislation. The desire of countries to financially exploit their wildlife resources under the CITES umbrella can be used to provide resources and expertise to undertake ranching and sustainable exploitation programs in native habitats, as well as incentives for habitat conservation and creation of protected areas. The CITES conferences offer an ideal forum for the exchange of information on endangered species management programs, as well as establishment of exchange programs for biological research, wildlife law, and trade regulation techniques.

These approaches to CITES implementation are quite different from those seen thus far. But they offer much more than trade controls ever can on their own. First, they provide a means to supplement the resources that a government devotes to CITES and wildlife management, while avoiding the perception of infringing on national sovereignty. Second, this sort of strategy emphasizes the building of conservation infrastructures within exporting countries, the only long-term hope for species conservation. Third, this strategy can benefit species that do not enter the international trade. Fourth, in contrast to trade controls, wildlife protection efforts need not rely on the coordinated functioning of a party's administrative, customs, and law enforcement infrastructures. Similarly, many of the benefits are not dependent on intricate international policy coordination and would accrue independently of actions taken in other countries.

The parties themselves have called for member states to include CITES technical assistance in development aid programs and to provide special funding and staff to the secretariat and developing countries. (See, for example, CITES Secretariat, 1982a.) It will, of course, be argued by some parties that this approach threatens their national sovereignty.[12] But a country is unlikely to be able to control its international wildlife trade according to the tenets of CITES if it is not able to manage its wildlife domestically. Placing artificial constraints on CITES' ability to improve the domestic capabilities and infrastructures necessary for proper CITES implementation is self-defeating. CITES is not "merely one tool." It is virtually the only active international legal tool available to achieve these ends. If pursued with sensitivity, the measures suggested here need not raise the sovereignty hackles of the benefiting countries.

Amidst the problems plaguing CITES, there are some signs of progress. CITES parties have begun to organize research into the biology and trade status of several species. Crocodile experts, in coordination with the tanning industry, are attempting to devise strategies by which the behavior of wildlife collectors in the field will be directed away from taking protected and undersized specimens (F. W. King, IUCN Species Survival Commission, pers. com., 1989). The CITES secretariat recently reached agreement with the government of Bolivia to revise its wildlife laws and establish priorities for ecological research (CITES Secretariat, 1988). These efforts constitute only a small start in the right direction and often focus on perfecting trade controls rather than building long-term management capabilities. Indeed, pursuit of the proposed strategy is not without problems. Such a change might be viewed by a significant portion of the NGO community as a sellout to wildlife trading interests. Moreover, the approach is far less symbolically satisfying than bringing several more hundred species under CITES "protection" every two years or "closing" loopholes in the convention through nonbinding resolutions. Funneling money, expertise, and other resources into CITES agencies at the national level, and into biological research, would also be quite expensive. But if the money can be found, progress can be made. As Mares and Ojeda (1984, p. 583) conclude: "We realize that education is a slow and tedious process offering no guarantee of success. But if our current efforts at dealing with species rarefaction continue on for the next fifty years, there will be few species left to preserve."

BEYOND THE RHETORIC

International application of the Endangered Species Act is on the wrong course. While conservationists are concerned that domestic implementation of the act has come face-to-face with perceived economic realities and has slowed to a plodding pace, international efforts are based on rhetoric rather than substance. Ironically, landmark provisions ensuring NGO participation in CITES affairs have given NGOs just as large a stake in the treaty's political and rhetorical development as that of governmental participants. As a result, rather than performing their indispensable watchdog function, NGOs have often contributed to the treaty's basic dysfunction. At present there is no framework for the future success of CITES trade controls. Policy initiatives in foreign aid and development assistance, potentially provided for by Sections 7 and 8 of the act, provide the real foundation for achievement of the act's goals abroad. CITES offers hidden yet powerful means to further these same goals by building up the wildlife management capabilities of member states. By complementing the provisions of Sections 7, 8, and 8A(e), CITES could make the Endangered Species Act much more effective internationally.

Notes

The views expressed in this essay are those of the authors and do not necessarily reflect the views of their employers, the World Resources Institute, or the Department of Justice.

1. Pub. L. 91-135, 83 Stat. 275, §2.
2. Pub. L. 91-135, 83 Stat. 275, §5(b).
3. ESA §7(a)(2), 16 U.S.C. §1636(a)(2).
4. 50 C.F.R. 402.02 (1988).
5. *Defenders of Wildlife v. Hodel*, 707 F. Supp. 1082 (D. Minn. Feb. 15, 1989), affirmed, No. 89-5192 (8th Circuit Aug. 10, 1990).
6. As of June 1989, there were 507 foreign species on the U.S. endangered species list. Although this is almost half the total number of 1,046 species on the list, it represents a fraction of the number of endangered species worldwide. As with domestic species, most of the foreign species listed were added in the early years of the act.
7. See the essay by Steven Yaffee on the consultation process in Part II of this volume.
8. For a bibliography covering regulation of the wildlife trade see Kosloff and Trexler (1987).
9. Approximately $2 million per year is spent on the FWS wildlife inspector program alone. Other U.S. CITES expenditures include undercover investigations by FWS special agents, activities conducted by the FWS's Division of Law Enforcement, litigation activities conducted by the Department of Justice and U.S. Attorneys' offices, and, of course, permit-related activities conducted by the FWS. Because of the overlap of CITES responsibilities with other legal requirements (such as the U.S. Department of Agriculture's

bird quarantine requirements), it is difficult to arrive at a precise estimate of resources devoted to CITES implementation.

10. At least one long-time CITES observer concedes that characterizing CITES as the most successful international wildlife instrument ever is premised on the failure of all prior instruments; G. Hemley, director of TRAFFIC (U.S.A.), pers. com.

11. One participant in the drafting process argues that he saw CITES as a mechanism by which to promote infrastructural development and information collection within developing countries. He assumed that such developments would be understood to be a prerequisite to successful CITES implementation and that they would be carried out; E. Baysinger, FWS, pers. com. Other long-time observers, however, have stressed that CITES is a narrow trade control instrument and nothing but that. A leading NGO observer, for example, has stated: "The Convention is a precise and finely defined legal instrument that acts at the point of exit and entry; it is not a policy statement on conservation nor is it the whole answer to our conservation problems. It is no substitute for habitat protection, but merely one tool that must be used in conjunction with other activities to help ensure a future in the wild for many important taxa. There is the definite need at regular intervals to clarify the point that this Convention deals with just the trade" (Lucas, 1979, p. 10).

12. As Canada and other parties have declared: "We are disturbed that over recent years more and more Parties and certainly most nongovernmental organizations are viewing CITES as an international agreement through which wild natural resources can be managed within a country as well as in international trade. Many recent proposals for inclusion of species in the appendices appear to be based upon that fact that the proposing State apparently does not like the way the species is being managed within the host nation. Such proposals also generally lack evidence indicating that the species is being illegally internationally traded. . . . We stress this is a trade agreement, and we should only be concerning ourselves with the regulation of international trade in the designated species" (CITES Secretariat, 1982b, p. 761).

References

Boardman, R. 1981. *International Organizations and the Conservation of Nature*. Bloomington: University of Indiana Press.

CITES Secretariat. 1982a. "Technical Cooperation." *Proceedings of the Third Meeting of the Conference of the Parties*. Conf. 3.4, p. 43.

CITES Secretariat. 1982b. "Ten Year Review of Appendices." *Proceedings of the Third Meeting of the Conference of the Parties.* Doc. 3.28, p. 761.

CITES Secretariat. 1988. "Interpretation and Implementation of the Convention: Review of Alleged Infractions." *Proceedings of the Sixth Meeting of the Conference of the Parties.* Doc. 6.19 (rev.), p. 541.

Kosloff, L. H. 1990. "Enforcement of the Convention on International Trade in Endangered Species of Wild Fauna and Flora in the United States." Master's thesis, University of California–Davis.

Kosloff, L. H., and M. C. Trexler. 1987. *The Wildlife Trade and CITES: An Annotated Bibliography for the Convention on International Trade in Endangered Species of Wild Fauna and Flora.* Washington, D.C.: World Wildlife Fund.

Lucas, G. 1979. Untitled article. *Threatened Plants Committee Newsletter* 4, July 1979, p. 10.

Lyster, S. 1985. *International Wildlife Law.* Cambridge: Grotius Publications, Ltd.

Mares, M. A., and R. A. Ojeda. 1984. "Faunal Commercialization and Conservation in South America." *BioScience* 34:584.

Nichol, J. 1987. *The Animal Smugglers and Other Wildlife Traders.* New York: Facts on File.

Trexler, M. C. 1990. "The Convention on International Trade in Endangered Species of Wild Fauna and Flora: Political or Conservation Success?" Ph.D. thesis, University of California–Berkeley.

Vermeij, G. J. 1986. "Biology of Human-Caused Extinction." Pp. 28–49 in B. G. Norton (ed.), *The Preservation of Species: The Value of Biological Diversity.* Princeton, N.J.: Princeton University Press.

World Wildlife Fund. 1985. "South America Faces Wildlife Trade Crisis." *WWF Focus*, March/April 1985, p. 1.

THE APPROPRIATIONS HISTORY

by

FAITH CAMPBELL

EFFECTIVE PROTECTION for endangered species depends on adequate financing. Funds are necessary to employ staff to evaluate whether species should be listed, to assess the possible impact of proposed activities, and to devise and carry out programs to restore species to viable population levels. In addition, law enforcement agents must be available to investigate violations of the law's legal protections. Finally, habitat areas must be purchased and managed to ensure that they remain suitable for species' use.

Several federal agencies are involved in protecting endangered species: the Fish and Wildlife Service (FWS) for terrestrial species, the National Marine Fisheries Service (NMFS) for marine species, and the U.S. Department of Agriculture (USDA) for enforcing restrictions on importation and exportation of listed plant species. The federal land management agencies, primarily the Forest Service and Bureau of Land Management (BLM), also have major responsibilities for protecting endangered species. The overriding theme of the appropriations history of the Endangered Species Act (ESA) is that none of these agencies has been adequately funded to carry out a comprehensive program of species protection.

RESOURCES VS. NEED: THE WIDENING GAP

Tables 2 and 3 show that although funding for endangered species conservation work has increased for all relevant agencies, it has not grown fast enough to keep up with even the tardy expansion of the federal endangered species list—much less the ever-expanding number of nonlisted species in need of protection.

Fish and Wildlife Service

Overall, FWS spending on endangered and threatened species, excluding land acquisition, has increased three times faster than inflation since the first year of the program. This is slightly faster than the pace of listings for native U.S. species. Much of this growth took place during the 1970s, however, when funding

TABLE 2

FUNDING FOR THE USFWS
ENDANGERED SPECIES PROGRAM
(in millions of dollars)

Fiscal Year	Funding
1974	4.657
1975	5.542
1976	9.486
1977	13.330
1978	16.534
1979	18.869
1980	20.087
1981	22.782
1982	17.769
1983	20.459
1984	22.205
1985	26.944
1986	28.824
1987	29.764
1988	31.066
1989	33.616

TABLE 3

**FUNDING FOR ENDANGERED SPECIES PRO-
GRAMS OF THE FOREST SERVICE AND BLM
(in millions of dollars)**

Fiscal Year	Forest Service	BLM
1980	NA	1.99
1981	1.544	1.44
1982	1.537	1.39
1983	1.31	1.65
1984	2.43	2.65
1985	2.46	3.95
1986	2.37	3.75
1987	3.62	3.39
1988	4.49	4.3
1989	7.15	5.0

outpaced listings. During the early years of the Reagan administration, spending declined severely. According to one source (Barton, 1987, p. 326), the endangered species program increased 3 percent in constant dollars during the Reagan years. In other words, funding in 1987 was almost the same as in 1981, despite the listing of 237 species during the period. Actually, the funding situation is even worse than these figures suggest. Since the 3 percent increase apparently includes land acquisition, it masks the decline in real resources allocated specifically for listing and conserving endangered species.

National Marine Fisheries Service

The NMFS does not separate funding for endangered species under its jurisdiction from that for species protected under the Marine Mammal Protection Act. Consequently, it is difficult to determine actual funding levels. In 1987 and 1988, funding was $3.168 million and $3.614 million, respectively. Appropriations have apparently never exceeded $4 million, a token sum. In contrast, the Center for Environmental Conservation has recommended an appropriation of $7.5 million. As we shall see, other federal agencies charged with species protection struggle with a similar gap between their resources and the magnitude of the problems they face.

Department of Agriculture

The U.S. Department of Agriculture is responsible for enforcing both the Endangered Species Act and the import and export regulations for plants called for in the Convention on International Trade in Endangered Species (CITES). There is no separate appropriation for the program, but USDA is assumed to use the amount authorized for this purpose ($1,850,000). This authorization represents 2 percent of the total appropriation for the USDA division carrying out these responsibilities. In 1985, the Endangered Species Act Reauthorization Coalition (ESARC—a group of more than thirty scientific, animal welfare, and conservation organizations) recommended increasing the authorization to $2.5 million to allow for inflation and to encourage the USDA to improve its monitoring of plant trade, training of port inspectors, and species identification materials. The need to hire botanists for two additional ports that receive large numbers of imported orchids was of particular concern at the time.

Federal Land Management Agencies

Under Section 7 of the act, the major federal land management agencies have a legal obligation to further the act's purposes by promoting recovery of listed species on their lands. Since a significant number of listed and candidate species occur on public lands (see Table 4), managing agencies have the opportunity to halt or even reverse population declines and perhaps obviate the need for listing. Currently, the primary need is to hire adequate numbers of biologists in field offices where land-use decisions are made. These biologists must be given sufficient authority to curtail land uses that are harmful to listed or candidate species. Additional funds are needed to erect and maintain fences and signs protecting fragile areas and to patrol areas subject to vandalism.

TABLE 4

**NUMBER OF LISTED SPECIES FOUND
ON FEDERAL LANDS**

Fiscal Year	Forest Service	BLM
1981	97	?
1982	99	75
1983	99	85
1984	110	96
1985	124	109
1986	132	127
1987	153	140

FOREST SERVICE

The number of listed species on Forest Service land grew by 58 percent between 1981 and 1987. During that time, funding for endangered species almost tripled as a result of congressional add-ons. Even so, the $4.49 million available for endangered species protection in fiscal year 1988 still falls short of the need. Virtually all of the funds have been spent on fewer than a dozen species of mammals and birds. Only when Congress began earmarking money for work with endangered plants did the Forest Service establish a comprehensive program for that kingdom. Recognizing the need to increase funding for endangered species, Forest Service staff are now developing a program to "sell" endangered species to higher levels of the service, its parent agency, the Department of Agriculture, and the president's Office of Management and Budget.

BUREAU OF LAND MANAGEMENT

The number of listed species on BLM lands grew by 87 percent between 1982 and 1987, while funding slightly more than doubled. Although more funds are needed, the 1988 allocation of $4.3 million does represent 23 percent of BLM's total wildlife budget. Thus endangered species funding cannot be increased until we overcome BLM's overall funding inadequacies. The BLM's budget has been cut nearly 10 percent since 1981. According to Barton (1987, p. 341), BLM's renewable resource programs have been hit particularly hard. Even though BLM manages more habitat than any other agency, its habitat management program has been cut 21 percent in constant dollars. This is the largest cut of any of the federal agencies with wildlife habitat responsibilities.

NATIONAL PARK SERVICE

It is more difficult to assess the National Park Service's funding needs for endangered species protection because its resource budget is compiled by individual park units. Nevertheless, it is clear that considerably more money is needed for at least some parks, especially those in Hawaii. Relatively few species listed as endangered or threatened are found in these parks because the FWS trusts the National Park Service to protect species whether they are listed or not. But inadequate funding and staff undercut many of the parks' efforts.

Haleakala National Park in Hawaii, particularly its Kipahulu District, is the highest-priority area in the entire national park system in terms of biological conservation work. The Kipahulu District, relatively pristine in the late 1960s, has since been invaded by feral pigs and goats that now threaten to destroy the area. To control these threats in the Kipahulu and Crater districts, Haleakala National Park requires a resource budget of about $700,000—more than three times the Bush administration's request for 1991.

Hawaii Volcanos National Park has advanced further in ungulate control, but its native vegetation is threatened by exotic plants. The park estimates that it will cost $900,000 per year to control pigs, goats, and invasive plants in a reasonable portion of the park's most pristine habitats—double the appropriations for 1990. To ensure long-term continuity, these funds must come from the park's permanent base funds, not annual add-ons or park entrance fees.

THE PRICE OF INADEQUATE FUNDING

Delays

The most conspicuous problem related to inadequate funding is long delays in the listing process. There are currently 3,900 candidate species for listing, of which nearly 1,000 are known to qualify for protection. Completion of the listing process and initiation of active protection and management, however, have been prevented by lack of staff. According to Defenders of Wildlife and other members of the ESARC Committee, listing these Category 1 candidates within ten years would require an annual appropriation of $7 million (in 1988 dollars)—double the 1990 appropriations. Carrying out status surveys on the 2,900 Category 2 candidates (about which less is known) within ten years would require an additional appropriation of $1.6 to $1.75 million per year. Once this is accomplished, listing and subsequent protection actions would require further funds.

Species Recovery

The ultimate goal of the Endangered Species Act is not to list species per se but to protect them so that they may recover and be taken off the threatened and endangered list. The recovery portion of the program should have expanded rapidly to support research and protection of the increasing number of species on the list. It has not. In its first years, the Reagan administration defended delays in listing additional species by saying that it wanted to concentrate on the recovery of species already on the list. Nevertheless, funding for recovery in 1988 was only 26 percent greater than in 1981 (in current, not constant, dollars)— far short of the 66 percent increase in U.S. species listed during this time.

The consequences of failing to fund the recovery program adequately show up in official statistics. Despite legal requirements, recovery plans have been completed for only 263 of 998 listed species, and approximately 40 percent of the 420 U.S. listed species still have no approved recovery plan. Furthermore, many plans are now out of date and implementation has lagged. Less than half of the existing plans are being actively implemented, and then only to the minimal extent needed to fend off extinction. The 1984 FWS recovery implementation report stated that only 23 species were known to be increasing, 10 were considered stable, the status of 154 was unknown, and over 30 species were at crisis population levels or presumed extinct.

Sections 6 and 7

Funding for Section 6 cooperative grants, under which the states are encouraged to protect federally listed species, has also failed to keep pace with the increase in either the number of species listed or the number of cooperative agreements signed. The current appropriation of $4.3 million is approximately one-quarter of the level needed to fund each agreement at the same level as in 1977 (excluding higher costs due to inflation and the increased number of listed species). Funding inadequacies plague Section 7, as well, under which all federal agencies must consult with the FWS or the NMFS if they wish to take action that may affect an endangered or threatened species or modify its critical habitat. If these consultations are to proceed promptly and still be sufficiently thorough to ensure protection, the FWS must have adequate staff to respond to agency contacts. Defenders of Wildlife point out that while the number of consultations begun each year is now about five times that in 1979, the funds appropriated are lower in real dollars.

Land Acquisition and Management

Another key area is habitat acquisition coupled with appropriate management. Over the past eight years, land acquisitions funded by the Land and Water Conservation Fund have continued to grow, although at a lower rate than in the 1970s (Barton, 1987, p. 326). These acquisitions were funded by Congress largely over administrative objections (see Table 5). Endangered species benefit from these acquisitions to some extent.

TABLE 5

FUNDING FOR LAND ACQUISITION BY USFWS
(in millions of dollars)

Fiscal Year	Request	Appropriation
1981	11.42	9.303
1982	1.139	16.489
1983	1.567	35.2
1984	0	46.297
1985	45.54	63.218
1986	1.5	38.7
1987	1.5	42.4
1988	1.6	51.754

William Reffalt, former director of the FWS Refuge Management Division, has estimated that in past years the proportion of acquisition projects (not funds) that included lands inhabited by endangered or threatened species was about 75 percent; in 1988, this figure fell to 50 percent.

Habitat acquisition, however, is not sufficient to ensure the survival of endangered species. Land acquisition without management is analogous to sailing a boat without a rudder. Although time and money are spent purchasing land for species protection, managers are left drifting without the means to chart a desired course. Thus, despite sizable increases for refuge operations, funding remains inadequate.

One of the most egregious examples of acquisition without proper maintenance is provided by the Hakalau Forest National Wildlife Refuge in Hawaii. Since 1984, Congress has provided $16.7 million above the administration's request for purchase of native forest in Hawaii to protect five endangered species of forest birds. Hakalau Refuge was purchased in October 1985. Additional funds were provided in 1988 to purchase Keauhou-Kilauea National Wildlife Refuge. Congress has also provided additional funds for maintenance of Hakalau and the other twelve refuges in the Hawaiian Islands complex.

Welcome as these funds were, they still fell short of the need. In 1988, maintenance funds for Hakalau and the twelve other refuges totaled $1,280,000, including a congressional add-on of $500,000. Hakalau Refuge's share was $170,000. This permitted the hiring of one supervisor provided with an office (located over two hours away from the refuge) and a vehicle; stationing of one temporary employee at the refuge to supervise fence construction and compliance with grazing permits; development of fire control and grazing management plans (but not their implementation); and a pilot study of reforestation methods.

Proper management of Hakalau required a budget of approximately $2.7 million for the next fiscal year (1989). The FWS needs to build 60 to 70 miles of fence to control inroads by cattle and feral pigs. The cost of building this fence has been variously estimated from $1 million to several times that figure, depending on the terrain encountered. To ensure that the terms of the current grazing leases are followed and to supervise the building and maintenance of fences, the refuge must have a permanent on-the-ground staff. It will cost approximately $700,000 per year to provide a minimal staff. Once fences are installed, cattle grazing should be ended and additional funds appropriated to maintain the fences, plant native tree species, remove pigs, and control banana poka (an invasive weedy vine). In all, annual maintenance budgets should be in the $1 million range. Without the necessary funds, the endangered bird species originally slated for protection are likely to be lost despite the best intentions behind the land purchase.

DIRECTIONS FOR THE FUTURE

The public is generally unaware of the disparity between the needs of the federal endangered species program and available resources. The endangered species program constitutes only 4 percent of the FWS budget—9 percent if one excludes land acquisition and permanent trust fund accounts (Barton, 1987, p. 329). Endangered species receive approximately 4 percent of the NMFS research and habitat conservation funds (Barton, 1987, p. 351), 9 percent of the Forest Service's wildlife habitat management budget, and 23 percent of the BLM's wildlife budget. To close the funding gap, several strategies must be pursued simultaneously.

Devoting More Existing Funds to Endangered Species

Although conservation organizations and the public apparently regard endangered species conservation as a high priority, this value is not reflected in funding. Game management still dominates the nation's wildlife conservation programs. During the Reagan years, while endangered species funding increased 3 percent in constant dollars, funding for the entire Fish and Wildlife Service increased by 8 percent (from $288 million in 1981 to $396 million in 1986), excluding permanent trust funds (Barton, 1987, p. 326). As noted above, endangered species benefit from the two programs that received the greatest increases: land acquisition and refuge maintenance. However, approximately 60 percent of the FWS's operating budget (about $240 million) goes to other programs such as habitat protection, research and development, migratory bird conservation, fishery enhancement, and wetland inventories. These funds come from general appropriations, not from taxes on hunting and fishing equipment or migratory bird hunting licenses. Rather than continuing the fiction that these needs are being met with funds earmarked for endangered species programs, the FWS could prioritize protecting endangered species under these programs.

As described earlier, a large proportion of Forest Service and National Marine Fisheries Service wildlife management dollars goes to nonendangered species. Much could be done if all the relevant agencies devoted a greater proportion of their wildlife budgets to endangered species. For this to happen, however, the whole spectrum of conservation organizations (ranging from hunter-oriented groups to the humane societies) will have to agree to reallocating funds away from well-established game programs. Of course, game management programs have their own funding mechanisms and well-entrenched bureaucratic and public support groups. Furthermore, certain game species do need improved management. Some duck species, for example, have declined precipitously in recent years. To justify dipping into other long-standing program funds or to obtain "new" money, the conservation community must agree on a source of dedicated tax funds to finance endangered species conservation.

Lobbying for Appropriations

The conservation community should also put more energy into lobbying for appropriations for wildlife management agencies. Conservation groups find it easier to lobby in opposition to weakening amendments or for reauthorization of the act than for the funds needed to implement programs. Within this lobbying effort, more attention must be given to the nuts-and-bolts aspects of management. As the situation at Hakalau Refuge illustrates, land acquisition is not enough.

Moreover, we should consider reducing the federal "matching share" for the Section 6 grants to state endangered species programs. Given the nation's budget deficit, it is unrealistic to expect the federal government to continue to pay 75 percent of program costs. It would be better to spread funds more broadly by reducing the federal share to 60 percent. This amount should still be sufficient to stimulate state efforts. In fact, it may be more effective since many states now feel that the chance of obtaining funds is so low that they do not even bother to apply.

Setting Priorities

Priorities must be set among species as well. From the standpoint of biological conservation, our efforts should be directed toward those communities or clusters of rare species that have a reasonable chance of survival. Acquisition and adequate management of refuges sheltering several species is generally more cost-efficient and ecologically sound than attempting to rescue single species. The few high-profile mammals and birds we have pushed to the very brink of extinction may be able to survive on voluntary contributions. Others, such as the bald eagle and peregrine falcon, are now able to recover on their own without further high-cost interventions.

Candidate species should also be listed in clusters rather than individually. For this to occur, however, the FWS and NMFS must overcome their current antipathy to listing and hire additional staff in the needed specializations (such as invertebrate zoologists and botanists).

Avoiding Last-Minute Rescues

Finally, the costs of mitigating the impact of development on endangered species could be cut substantially if the responsible agencies faithfully carried out their legal obligation to ensure that their actions do not harm such species. Although there would still be costs associated with placing adequate numbers of biologists in Forest Service and BLM district offices, the need for last-minute rescues would be reduced.

Reference

Barton, K. 1987. "Federal Fish and Wildlife Agency Budgets." Pp. 321–354 in Roger Di Silvestro (ed.), *The Audubon Wildlife Report, 1987.* Orlando: Academic Press.

IMPLEMENTING RECOVERY POLICY: LEARNING AS WE GO?

by

TIM CLARK AND ANN HARVEY

T HOSE INVOLVED in endangered species recovery programs often face extremely complex situations as they tackle the nuts-and-bolts work of saving species. Recovery programs over the last fifteen years have had to deal with technically demanding biological tasks and uncertainties, limited resources, numerous participants, and intense public scrutiny and involvement. Professionals working in these programs often view recovery primarily as a biological problem. Instead of paying explicit attention to policy and organization in recovery programs, they tend to attribute problems to bad luck, lack of resources, "politics," or lack of commitment in other organizations. Yet the organizational arrangements, decision-making processes, and other policy variables affecting recovery programs can be as critical to success as technical and biological tools. A better understanding of the policy and organizational dimensions of endangered species work could greatly enhance the effectiveness of many recovery programs.

Participants in recovery programs often view the problems they encounter as unique to their species and their program. But problems stemming from inappropriate organizational and decision-making arrangements may be more prevalent than is currently recognized. By looking at these programs from a policy and organizational perspective, one may detect common patterns that would otherwise remain masked. Ignoring these aspects of recovery can result in ineffective programs and even in species extinction. With so much at stake, it is imperative to develop a framework for analysis and to learn from past and present recovery efforts.

Notable successes have been achieved in many recovery programs. The American alligator, the bald eagle, and the peregrine falcon have recovered in many parts of their former ranges as a result of federal and state protection under the Endangered Species Act (ESA). Yet many accounts of recovery programs refer to implementation difficulties encountered by participants (Duff, 1976; Carr, 1986; Askins, 1987). Here we discuss four common features of recovery programs that have led to implementation problems. First, species recovery is a tremendously complex task involving numerous people who must somehow integrate their diverse perspectives into a workable program. Second, these people often have conflicting goals, some of which have more to do with controlling the project than saving the species. Third, rarely is there explicit consideration of organizational structures appropriate to the task of saving species; recovery programs tend to develop into traditional hierarchical bureaucracies. Fourth, intelligence failures and program delays often occur because of preconceptions held by decision makers and the large number of clearances required in programs with multiple participants.

To illustrate these points, we turn to the black-footed ferret recovery effort, which has received much attention. (Although the events described here are recent, already there are different interpretations of what occurred. For a different perspective see Thorne and Williams, 1988.) Even though we focus on the ferret recovery effort from 1981 to 1986—from the discovery of the Meeteetse population until its extinction in the wild—the four implementation themes addressed here were apparent throughout the past fifteen years. Our use of the ferret case history could be misunderstood as unduly critical—in fact, we have been urged to forget past implementation mistakes. But unless these persistent features of implementation are scrutinized and given some meaning through a policy and organizational framework, they will never be recognized for what they are and managed effectively. By using the ferret example we are not implying that it is an especially good or bad program. Rather, we suggest that the examples indicate the implementation problems found in many recovery programs and that the lessons to be learned from examining them can be useful in many other cases.

THE BLACK-FOOTED FERRET STORY

The black-footed ferret is one of the most critically endangered mammals in North America. It was listed in the U.S. Fish and Wildlife Service's Redbook of Endangered Species in 1964, and in 1976 it was placed on the Fish and Wildlife Service's endangered species priorities list. It is a solitary, nocturnal carnivore preying almost exclusively on prairie dogs. The ferret spends almost all of its time below ground in prairie dog burrows where it hunts and finds shelter. In 1920, an estimated 1 million ferrets existed in 40 million hectares of habitat (prairie dog colonies) over twelve states and two Canadian provinces (Anderson and others, 1986). Widespread prairie dog poisoning programs, with the goal of rangeland improvement, destroyed ferret habitat. This loss, combined with other factors such as disease, pushed ferrets to the very edge of extinction by 1980. In fact, many experts considered the ferret extinct by that time. In the 137 years since the ferret's scientific discovery, only two small populations have ever been studied—one in South Dakota (1964–1974) and the second near Meeteetse, Wyoming (1981–1987). Both wild populations are now extinct.

The Meeteetse ferrets were discovered serendipitously. When a ranch dog happened to kill a dispersing male, the source population of ferrets was found nearby occupying thirty-seven prairie dog colonies (about 3,000 hectares) scattered over about 260 square kilometers on nine ranches in a mix of private and public lands. The presence of this ferret population surprised everyone. A few months after the discovery, the U.S. Fish and Wildlife Service (FWS) transferred lead authority for the ferret recovery program to the Wyoming Game and Fish Department. Ferret ecology and behavior were studied extensively as ferrets were observed directly, tracked in snow, and radio-collared. Spotlight surveys each summer revealed peak annual numbers (129 ferrets including 25 litters in 1984). Although annual ferret losses were high, about 50 to 90 percent (Forrest and others, 1988), counts in early July 1985 showed a much lower population than in all previous years (58 including 13 litters). By early September, mark/recapture population estimates showed that the population had declined to 31, plus or minus 8 ferrets. By early October, the population had declined to 16, plus or minus 5. By November, only about 6 ferrets were thought to remain in the wild. The catastrophic loss of some 150 ferrets between fall 1984 and fall 1985 was documented. From July to September 1985, ferrets were lost at the rate of one every two to three days. Biologists attributed the decline to canine distemper, a disease that is 100 percent fatal to ferrets. Although techniques were developed to locate ferrets and extensive searches were conducted over several states, no ferrets were found.

During fall 1985, six Meeteetse ferrets were captured to prevent loss of the species. These ferrets were housed in close proximity, however, and two ferrets infected with canine distemper transmitted it to the other four. All six died shortly thereafter. Another six were hastily captured and housed individually; all survived. These six, added to the four thought to exist in the wild, constituted the world's known population—about ten individuals in early 1986. In 1986, the six captive ferrets did not reproduce, but the four wild ferrets produced ten young in two litters and most were added to the captive population. This brought the world's known population to eighteen, all in captivity. The captive ferrets produced seven surviving young in two litters in 1987. No more wild ferrets were found. Breeding success was better in 1988, with forty-four young in thirteen litters being produced. Ten of the forty-four young born in 1988 died. Reproduction was relatively high in 1989 and 1990. The fate of the species now depends on the more than 180 captive ferrets and any wild ferrets that may exist (Maguire and others, 1988).

The captive population, upon which the fate of the entire species appears to rest, is presently held in three locations in Wyoming, Nebraska, and Virginia. The agencies responsible for the ferrets are planning to divide the population further in order to minimize the chances of extinction by disease or some other catastrophe (Oakleaf, 1988). An interstate coordinating committee has been formed to study potential reintroduction sites (Thorne, 1988).

IMPLEMENTATION PROBLEMS

Complexity of Cooperation

Like most endangered species programs, the ferret program includes a number of governmental and nongovernmental participants who became involved, formally and informally, for a variety of reasons. More than twenty organizations and a hundred people have participated in the ferret program since 1981. The primary participants are the U.S. Fish and Wildlife Service, the Wyoming Game and Fish Department, ranchers, and the conservation community. The complexity of coordinating the actions of multiple participants in wildlife programs can compound an already difficult biological task. This is not to argue that participation should be limited to only a few. To the contrary, a diversity of participants ensures a diversity of knowledge, skills, and perspectives as well as a useful system of checks and balances that contribute significantly to recovery. But to blend all these interests and skills into coordinated action requires a carefully designed program and a clearly defined decision-making process.

Each participating organization in a recovery program has a distinct perspective from which it sees the program, its operation, and other parties. Each organization may differ from the others in its sense of urgency and its thoughts about the best location and means for recovery. In the ferret case, conflict arose over the question of when and where to initiate a captive breeding program. Because perspectives vary so much, the participating organizations may measure program success by entirely different criteria. Some agencies gauge success primarily by increases in a species' numbers, successful captive breeding, or gains in data collection leading to better understanding of the species' ecological requirements. For others, the major criterion is the degree to which they can prevent public controversy or control key aspects of the program. Disagreement over these criteria has led to conflicts in recovery programs as technicians, scientists, managers, and administrators seek to impose their readings of the "facts" and their values on other parties. (See Latour, 1987.)

Goal Displacement

It is probably safe to assume that all participants in endangered species programs genuinely seek species recovery. Yet despite this common goal, participants often disagree about the means to achieve it. Conflicts may arise over professional disputes, legal and procedural differences, differences of opinion on leadership and authority, and direct incompatibility of the suggested actions with other goals held by their organization. (See Pressman and Wildavsky, 1973.) Participants may try to redefine the recovery program to fit their own agency's priorities, which may be quite inflexible (Yaffee, 1982).

In some cases, an obvious conflict arises between the "task goal" (saving the species) and the "power goal" (gaining control of the program). "Goal displacement" occurs when an agency becomes more focused on power goals than on the biological task goals. A program driven by power goals is likely to compromise the biological task goals when the two come into conflict—as they invariably will. If the organization relies on a bureaucratic top-down style of decision making, power goals tend to dominate. If goals are set from the bottom up, however, by those most directly in contact with the species, task goals tend to dominate (Daft, 1983).

Friction between task and power goals was evident during the early years of the Wyoming ferret recovery program. The Wyoming Game and Fish Department, which had been given lead agency status by the U.S. Fish and Wildlife Service, wanted to keep the ferrets within Wyoming and carry out captive breeding only after the state had developed facilities to do so. David Weinberg (1986, p. 65) writes: "As [Wyoming] officials acknowledge, they never seriously considered allowing ferrets to leave the state [for captive breeding]. . . . We'd have no control over them." Several inquiries suggest that Wyoming's insistence on controlling the ferret program created unproductive conflict and caused delays (Carr, 1986; May, 1986).

Organizational Structures

One major cause of a program's failure to meet its goals is use of inappropriate organizational structures. Since most recovery challenges go well beyond the boundaries of a single organization, coalitions are formed that must integrate diverse structures, ideologies, and standard operating procedures. But agencies setting up a new recovery program rarely give explicit thought to how the recovery coalition should be structured. Programs are often set up along standard bureaucratic lines— not because this arrangement has proved most effective but because no other structure is considered. This in turn limits the set of solutions that seem plausible. In the first fifteen months of the ferret recovery program, the coalition's organizational structure evolved from a simple matrix to a traditional bureaucratic arrangement, where it has remained (see Figure 2).

Organizational structure has profound effects on the allocation of tasks and resources, the distribution of information, and the overall effectiveness of a program. Bureaucratization is a root cause of many implementation problems (Warwick, 1975). Those who implement recovery programs should give special consideration to other organizational structures, such as horizontally coordinated task forces and project teams (Clark and Cragun, in press).

A program's structure is both a determinant and an outcome of organizational power. A structure that concentrates decision-making authority in the hands of one agency makes it easy for that agency to downplay the role of other organizations and control information for its own benefit. The lead agency in the ferret program used several widely recognized bureaucratic mechanisms to consolidate its power. For example, it filled positions of power in its "advisory team" with its own personnel. By restricting permits and limiting contact with the press, it also controlled information. The bureaucratic structure chosen by Wyoming helped to solidify its top-down control over decision making, allocation of resources, definition of roles, and the timing and location of recovery activities. Unfortunately, this structure also closed the decision-making process to outside information and reduced the program's ability to be creative and responsive (see Etheredge, 1985).

Intelligence Failures

Intelligence failures and delays are common problems in recovery programs resulting in part from conflicts among par-

F IGURE 2

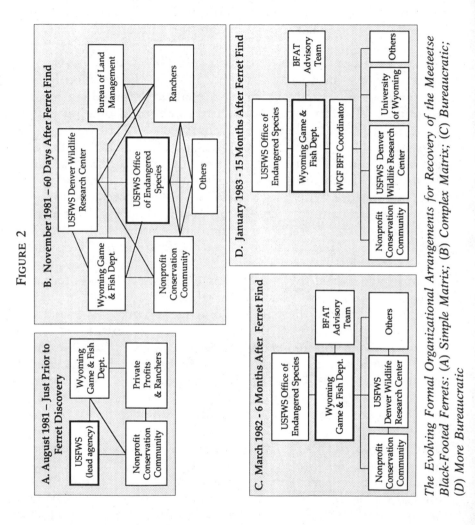

The Evolving Formal Organizational Arrangements for Recovery of the Meeteetse Black-Footed Ferrets: (A) Simple Matrix; (B) Complex Matrix; (C) Bureaucratic; (D) More Bureaucratic

ticipants, goal displacement, and use of inappropriate organizational structure. Wise decision making depends on
intelligence—the use of information or the acquisition, analysis,
and appreciation of relevant data. Even when information is
available to decision makers, they may dismiss it as erroneous,
inaccurate, or misleading. In the ferret program, agency officials
at first discounted 1985 field data indicating that the ferret
population was on a rapid decline. Officials took the most san-
guine view of the situation, arguing that it was just a normal
population fluctuation, that the field methods and data were in
error, or that the ferrets had migrated elsewhere (Weinberg,
1986; Randall, 1986; Zimmerman, 1986).

A root cause of intelligence failures, according to Betts (1978),
is that decision makers operate under premises that constrict
their perceptions and lead to "selective inattention" or outright
blindness to facts and new ideas. These preconceptions can
block learning, change, and adaptation. And organizational arrangements that stifle legitimate dissent exacerbate intelligence
failures.

In a difficult task like recovering species where numerous
parties are involved, one must expect disagreements over the
best course of action. When dealt with constructively, such conflicts have provided alternative solutions for the group to consider. But poorly managed conflict can cause substantial delays
that reduce the chances of successful recovery. Because Wyoming had neither a captive breeding facility nor the resources or
staff to build one—and because of their agency's strong opposition to sending ferrets outside Wyoming—captive breeding of
ferrets could not begin at once. It took about two years of extensive bargaining between Wyoming and other parties and the dramatic collapse of the wild population before Wyoming initiated
captive breeding in late 1985 (Weinberg, 1986; Randall, 1986).

Program delays are often hard to separate from program failures. Does Wyoming's move to breed ferrets in captivity—which
began a year later than recommended by field teams and conservationists (Weinberg, 1986) and after the wild population had
sharply declined—count as a failure or a success? In view of the
captive breeding program's results in 1988, some observers may
reasonably argue "better late than never." While this may be
true, the delicate nature of a species recovery program and the

irreversible consequences of failure behoove us to work toward improved conflict management so that delays are reduced.

IMPROVEMENTS

How can participants in recovery programs begin to deal with these implementation problems? To improve future performance in conserving species and the ecosystems on which they depend, one must recognize the complexity of the work to be done. This means developing a broad understanding of the interactive web of biological, organizational, and policy components. Such a "systems perspective" is very different from the conventional views held by traditional biologists and bureaucrats—views rooted in strict university disciplines and reinforced in agency cultures and loyalties (Brewer, 1988).

Apart from advancing the technical aspects of species recovery, improvement in recovery programs is possible in three areas: policies, organizations, and individuals. The ideas presented below suggest problem-solving techniques that can broaden participants' perspectives and improve their ability to adapt quickly to the demands of species recovery. Many recovery programs face extremely tight budgets, of course, and participants may view some of these suggestions as too time-consuming and expensive to be practicable. Yet we believe that these ideas can help recovery programs avoid common pitfalls that have hindered effective action in the past. Since we can offer only the briefest introduction to these ideas here, we urge readers to delve into the references we cite.

Improving the Process

In defining a recovery challenge, it is essential to explore its history, scientific and management context, and trends and to identify all factors that may have a bearing on the program's success. Some of these factors, particularly policy and organizational variables, are often neglected. Organization and management structures, resource limitations, uncertainty, and jurisdictional issues are just a few of the variables that can affect decisions and ultimately the outcome of the program. Many of these variables involve participants' values. The policy sciences' problem-solving tools are specifically designed to address both technical issues and values. In contrast to technical experts who generate basic knowledge, policy scientists look at how knowledge is used and, simultaneously, at how well the process itself is working.

One model that could be very useful for recovery programs is the decision seminar—a technique designed to allow a group of specialists and decision makers to integrate their knowledge to solve complex problems (Lasswell, 1960; Brewer, 1975). A core group of ten to fifteen participants must be willing to commit the time needed to understand the problem (over months or years if necessary), although the seminar is also open to outsiders. The group maps the context of the problem, determines its past trends and probable outcomes, and decides which options are available. The process by which decisions are made is also considered. Participants' independent assessments of the problem are compared, common views are discussed, and discrepancies are considered. All relevant methods for analyzing the problem are used, and new methods are encouraged. When the group arrives at a decision, responsibilities for carrying it out are assigned. Documentation of participants' activities becomes the group's "institutional memory" (Brewer, 1975). An interdisciplinary approach is essential. Many recovery programs incorporate certain aspects of the decision seminar model. But, for the most part, they lack the explicit attention to multiple methods and the breadth of analysis that characterize decision seminars. The participation of the IUCN Captive Breeding Specialist Group (CBSG) in the ferret recovery program since 1985 has improved the program's technical capabilities and broadened discussion of a range of ideas and problem-solving approaches. This has brought the program a little closer to the decision seminar model. Although the program still functions under several policy and organizational constraints dictated by Wyoming, CBSG's participation to date has resulted in a more focused problem-solving orientation and has contributed greatly to the success of the captive breeding effort.

Another tool that has proved useful in species recovery programs is decision analysis. Decision analysis allows managers to integrate ecological theory, objective data, subjective judgments, and financial concerns in making decisions when conditions are uncertain (Maguire, 1986). Probabilistic models are developed relating the outcomes of alternative actions to random events in the environment, and probability values are assigned to each possible outcome of a decision. For example, the probability of species extinction can be estimated under current management conditions and then under different management scenarios. The probabilities and effects of random events such as severe weather and disease, as well as the cost of different management actions, can be explicitly considered. Parties that dispute the facts can see where they agree and disagree and suggest ways of assembling information to resolve disputes. Analysts have applied decision analysis to conservation programs for the critically endangered Sumatran rhino and other species (Maguire and others, 1987, 1988).

Adaptive management (Hollings, 1978) is a third way of guiding a recovery group. In this approach, decision making is treated explicitly as a process of making mistakes and correcting errors (Brewer, 1988). Instead of seeking a single "best answer," managers should consider many plausible solutions, adapting to changes in the problem and its context. The key to adaptive management is to monitor the outcomes of decisions in order to cut one's losses when "solutions" are not working. Since recovery programs almost always involve risk and uncertainty, managers should use contingency planning to anticipate the possibility of failure.

Improving the Organization

Organizations are more than a collection of people. They have established norms, traditions, and activities above and beyond the individuals who direct and staff them. The complex, rapidly changing, and highly uncertain nature of endangered species recovery programs requires organizational arrangements that fit these characteristics. Since highly bureaucratized organizations with rigid standard operating procedures usually lack the flexibility needed, managers should question whether the program's organizational structure is hindering the recovery effort.

An effective organization should process information well and learn rapidly from its own mistakes. Useful models for endangered species recovery include task forces and project teams operating under adaptive management and decision seminar guidelines. Task forces tackle temporary problems; project teams address problems that need long-term, continuous coordination (Daft, 1983). A recovery team should be composed of professionals with formal training and experience, experts who can focus on completing the job successfully and are willing to accept the uncertainty inherent in such challenges.

Certain characteristics are essential. As the recovery task changes, the team must be able to respond quickly. Communication practices that facilitate high creativity—such as emotional supportiveness, brainstorming, and objective evaluation of ideas—are helpful. A willingness to examine every alternative is essential. Teams must avoid groupthink in which disagreements are muted in the interest of maintaining group cohesion (Janis, 1972). A strong, mutually supportive atmosphere in which mistakes will not result in withdrawal of the group's support is important. Mistakes should be viewed as occasions for improving the system.

Clark and Cragun (in press) provide a framework for analyzing organizational problems and for implementing change in species recovery programs. This fourteen-step procedure includes four major stages: defining the problem, developing alternative strategies, developing an action plan, and implementing and evaluating the action plan. It can guide participants in defining problems and objectives, identifying forces that could help or hinder movement toward objectives, analyzing strategies to overcome obstacles, outlining specific tasks to be accomplished, and evaluating the success of their efforts. In short, it provides an explicit method for solving both technical and organizational problems.

Improving the People

Many participants and observers believe that the root cause of faltering programs is misguided or selfish people. Although this "human relations" view of organizations oversimplifies the many complex organization, management, and policy aspects introduced here (see Hall, 1987; Ham and Hill, 1987), improvements can also occur at the individual level. Individuals are molded by conventional experience, established policy prescriptions, and agency structures and procedures. Nevertheless, individual performance in a recovery program is an important key to success. And it can, in many cases, be improved.

If only people acted with more professional integrity, one often hears, programs could be improved. But as Richard Betts (1978, p. 82) notes, "Integrity untinged by political sensitivity courts professional suicide." Betts suggests that individuals can try to improve programs by asking hard questions of their superiors, acting as socratic agnostics, nagging decision makers into awareness of the full range of uncertainty, and making authorities' calculations harder rather than easier. Of course, most leaders will not appreciate this approach. Simply providing more reliable facts or new arguments to decision makers will not reverse their basic beliefs. Analysis is often less important than values and preconceptions as a basis for decision making (Betts, 1978). Real solutions depend on the openness of decision makers and their understanding of the premises they use in accepting or rejecting information and ideas.

The sheer complexity of endangered species and ecosystem conservation tells us there is no single, straightforward, technocratic recipe for success. The essential challenge in species and ecosystem conservation, as in all complex situations, has always been to address unbounded problems successfully even though our analytical resources are bounded. Real improvements will come about by refining the conceptual tools that enhance our understanding of complex conservation problems and by developing practical tools that allow us to deal with the problems realistically. A number of conceptual and practical tools already exist but go largely unused. Improvements will not come quickly, even with increased use of these tools. There are many barriers to learning and improvement, but with so much at stake we must learn to recognize and overcome them.

Note

We would like to thank Thomas M. Campbell III, Leonard Carlman, Archie Carr III, Denise Casey, Pamela Parker, Debra Patla, Chris Servheen, John Weaver, Michael B. Whitfield, Dusty Zaunbrecher, and three anonymous reviewers for their thoughtful comments on earlier drafts of this essay. Not all the reviewers have endorsed its content, but such is the nature of the process in which we all work in trying to protect and restore endangered species.

References

Anderson, E., S. C. Forrest, T. W. Clark, and L. Richardson. 1986. "Paleobiology, Biogeography, and Systematics of the Black-Footed Ferret, *Mustela nigripes* (Audubon and Bachman), 1851." *Great Basin Naturalist Memoirs* 8:11–62.

Ascher, W. 1986. "The Evolution of the Policy Sciences: Understanding the Rise and Avoiding the Fall." *Journal of Policy Analysis and Management* 5:367–373.

Askins, R. 1987. "Wolf Recovery in the Yellowstone Ecosystem: A Progress Report." *Endangered Species Update* 4 (11):1–4.

Betts, R. K. 1978. "Analysis, War, and Decision: Why Intelligence Failures Are Inevitable." *World Politics* 31:61–89.

Brewer, G. D. 1975. "Dealing with Complex Social Problems: The Potential of the 'Decision Seminar.'" In G. D. Brewer and R. D. Brunner (eds.), *Political Development and Change: A Policy Approach*. New York: Free Press.

Brewer, G. D. 1988. "An Ocean Sciences Agenda for the 1990s." McKernan Lecture, University of Washington, Seattle, February 1988.

Carr, A. 1986. "Introduction: The Black-Footed Ferret." *Great Basin Naturalist Memoirs* 8:1–7.

Clark, T. W. 1986. "Professional Excellence in Wildlife and Natural Resource Organizations." *Renewable Resources Journal*, Summer, pp. 8–13.

Clark, T. W. and J. Cragun. In press. "Organization and Management of Endangered Species Programs." In B. A. Wilcox, B. Marcot, and P. F. Brussard (eds.), *The Management of Viable Populations: Theory, Application, and Case Studies*. Palo Alto: Center for Conservation Biology, Stanford University.

Craighead, F. C. 1979. *Track of the Grizzly*. San Francisco: Sierra Club Books.

Daft, R. L. 1983. *Organization Theory and Design*. St. Paul: West.

Duff, D. 1976. "'Managing' the Pupfish Proves Fruitless." *Defenders* 51:119–120.

Endangered Species Technical Bulletin. 1985. "Final Action Taken on Eight Species." 10(7):1, 6–8.

Etheredge, L. S. 1985. *Can Governments Learn?* New York: Pergamon Press.

Forrest, S. C., D. E. Biggins, L. Richardson, T. W. Clark, T. M. Campbell III, K. A. Fagerstone, and E. T. Thorne. 1988. "Black-Footed Ferret *(Mustela nigripes)* Population Attributes at Meeteetse, Wyoming, 1981–1985." *Journal of Mammalogy* 69:261–273.

Hall, R. H. 1987. *Organizations: Structure, Processes, and Outcomes.* 4th ed. Englewood Cliffs, N.J.: Prentice-Hall.

Ham, C., and M. Hill. 1987. *The Policy Process in the Modern Capitalist State.* Sussex, England: Wheatsheaf Books.

Harvey, A. H. 1987. "Interagency Conflict and Coordination in Wildlife Management: A Case Study." Master's thesis, University of Michigan.

Hollings, C. S. (ed.). 1978. *Adaptive Environmental Assessment and Management.* New York: Wiley.

Hornocker, M. 1982. Letter to the editor. *Wildlifer,* Nov.–Dec. pp. 51–52.

Janis, I. L. 1972. *Victims of Groupthink: A Psychological Study of Foreign-Policy Decisions and Fiascoes.* Boston: Houghton Mifflin.

Lasswell, H. D. 1960. "Technique of Decision Seminars." *Midwest Journal of Political Science* 4:213–236.

Lasswell, H. D. 1971. *A Pre-view of the Policy Sciences.* New York: American Elsevier.

Latour, B. 1987. *Science in Action.* Cambridge: Harvard University Press.

Maguire, L. A. 1986. "Using Decision Analysis to Manage Endangered Species Populations." *Journal of Environmental Management* 22:345–360.

Maguire, L. A., U. S. Seal, and P. F. Brussard. 1987. "Managing Critically Endangered Species: The Sumatran Rhino as a Case Study." Pp. 141–58. In Michael Soulé (ed.), *Viable Populations for Conservation.* Cambridge: Cambridge University Press.

Maguire, L. A., T. W. Clark, R. Crete, J. Cada, C. Groves, M. L. Shaffer, and U. S. Seal. 1988. "Black-Footed Ferret Recovery in Montana: A Decision Analysis." *Wildlife Society Bulletin* 16:111–120.

May, R. M. 1986. "The Black-Footed Ferret: A Cautionary Tale." *Nature* 320:13–14.

Oakleaf, B. 1988. "Divide the Breeding Colony? An Agreement Is Reached." *Black-Footed Ferret Newsletter* 5(1):2.

Pressman, J. L., and A. Wildavsky. 1973. "The Complexity of Joint Action." Pp. 87–124 in their *Implementation*. Berkeley: University of California Press.

Randall, P. 1986. "Survival Crisis at Meeteetse." *Defenders* 61(1):4–10.

Salancik, G. R., and J. Pfeffer. 1977. "Who Gets Power—and How They Hold on to It: A Strategic-Contingency Model of Power." *Organizational Dynamics*, Winter, pp. 2–21.

Schön, D. A. 1983. *The Reflective Practitioner: How Professionals Think in Action*. New York: Basic Books.

Thorne, E. T. 1988. "1987 Black-Footed Ferret Captive Breeding Season Successful." *Conservation Biology* 2(1):11–12.

Thorne, E. T., and E. Williams. 1988. "Disease and Endangered Species: The Black-Footed Ferret as a Recent Example." *Conservation Biology* 2(1):66–73.

Warwick, D. P. 1975. *A Theory of Public Bureaucracy: Politics, Personality, and Organization in the State Department*. Cambridge: Harvard University Press.

Weinberg, D. 1986. "Decline and Fall of the Black-Footed Ferret." *Natural History Magazine* 95:62–69.

Yaffee, S. L. 1982. *Prohibitive Policy: Implementing the Federal Endangered Species Act*. Cambridge: MIT Press.

Zimmerman, D. L. 1986. "Protecting the Black-Footed Ferret." *Newsday*, Jan. 21, p. 13.

PART III

IMPLEMENTATION CHALLENGES

What an appalling indictment it is, what a disgrace to mankind, that the road to his so-called civilization should be built on the memories of extinct species and species on the way to extinction.

—THE RIGHT HONORABLE
EARL OF JERSEY

The "control of nature" is a phrase conceived in arrogance, born of the Neanderthal age of biology and philosophy, when it was supposed that nature exists for the convenience of man. The concepts and practices of applied entomology for the most part date from that Stone Age of science. It is our alarming misfortune that so primitive a science has armed itself with the most modern and terrible weapons, and that in turning them against the insects it has also turned them against the earth.

—RACHEL CARSON,
Silent Spring

WESTERN WATER RIGHTS
AND THE ACT

by

A. DAN TARLOCK

IF THE SCIENTIFIC JUSTIFICATION for the Endangered Species Act (ESA) is correct, we should be moving toward institutions that practice integrated ecosystem management on a wide geographic scale instead of concentrating our efforts on the protection of an arbitrary and limited number of species and their habitats. But we are not sure what integrated ecosystem management means, and we are reluctant to make major institutional changes to manage our resources toward this end. Nevertheless, there seems to be a positive and growing movement to preserve biological diversity. Although to date the Endangered Species Act still relies primarily on protection of individual species to preserve biodiversity, the act's mandate incorporates habitat preservation strategies as well. These strategies are less pronounced than those protecting national park and wilderness systems, but habitat protection is becoming a significant constraint on many development activities.

While habitat preservation may be the most effective remedy to our ever-expanding endangered species list, it is also the most far-reaching. The shift toward habitat preservation to protect endangered species—even on a modest scale—may significantly affect the operation of many present and future water projects. Water development activities are now evaluated, in part, by backdoor water-related federal land-use planning processes under the environmental programs that Congress has superimposed on resource development programs. The integration into western water law of the values represented by the Endangered Species Act will not be easy. But integration is necessary because the stakes are so high. The first step is to recognize that it is legitimate for the federal government to use the ESA to claim water rights.

The ongoing controversy over the operation of Glen Canyon Dam illustrates both the problems and the potential of the act. In the 1950s, the Upper Colorado River Basin states persuaded Congress to construct Glen Canyon Dam (upstream from the eastern entrance to the Grand Canyon National Park) to meet their 1922 Colorado River Compact water delivery obligations to the lower basin states. Glen Canyon Dam was part of a deal struck between the Sierra Club and the Bureau of Reclamation to keep a dam, which threatened Dinosaur National Monument, off the Green River at Echo Park. At the time, only a few environmentalists mourned the loss of remote and unused Glen Canyon and the movement went on to other issues. The dam filled in 1980 and began to change the ecology of the river: Cold water has been substituted for warm water because the releases are drawn from lower levels of the dam; downstream sediment transport necessary for beach replenishment has been greatly reduced; and the safety and enjoyment of the white water rafting industry may have been hurt by the pulsating hydroelectrical power releases from what is known as a "cash register dam."

Adverse downstream effects caused by large multipurpose dams are familiar to environmentalists. What is interesting about Glen Canyon Dam is that since 1982 the Bureau of Reclamation has allocated over $8 million from power revenues for a series of continuing environmental studies about the dam's effect on the river and its riparian environment. These studies were driven by the bureau's fear that environmentalists will bring a lawsuit to force the bureau to prepare an environmental impact statement (EIS) to assess the full effects of Glen Canyon's operations. The triggering event for such a suit is the "uprating" of the dam's generators. A principal focus of the EIS will be the endangered humpbacked chub. The National Environmental Policy Act (NEPA) does not apply to ongoing projects, but uprating the dam's generators could be classified as a new activity.

After several years of sidestepping the issue, on July 27, 1989, Secretary of the Interior Manuel Lujan announced that the Bureau of Reclamation would prepare an EIS on the operation of Glen Canyon Dam.

One possible consequence of the studies is a change in the operating regime of the dam. Smoother, warmer, and seasonably timed releases from the upper strata of the reservoir would benefit the humpbacked chub. These changes will not come easy, however. The Western Area Power Administration has taken the position that power generation has priority over all other uses. Thus it can operate the dam as a cash register. Upper basin states begrudge any release beyond what is necessary to fulfill their compact obligations to the lower basin states, and they relentlessly monitor the day-to-day operation of the dam. All traditional stakeholders in the "law of the river," however, are especially troubled by the Endangered Species Act. The act has the potential to trump all existing allocations and to subordinate all water rights to a judicially mandated flow regime. How, if at all, the EIS affects the traditional biases of the Bureau of Reclamation toward power generation will be a major test of the ESA and more generally the whole idea of ecosystem management.

A NEW WATER RIGHT?

When the major environmental regulatory programs of the 1970s were put in place, Congress gave little thought to the relationship between these programs and the existing resource allocation regimes primarily administered by the states. The Endangered Species Act is a prime example of what might be termed a "legislative overlay approach." Western states soon recognized that environmental regulatory programs had the potential to mandate flow patterns at variance with traditional allocation patterns—which until recently excluded instream flow maintenance. This potential crystallized after the Tellico Dam decision in which the Supreme Court held that species protection was an absolute federal duty once a finding of jeopardy was made.[1]

As a result of *TVA v. Hill*, the Endangered Species Act has effectively created de facto regulatory water rights. That is, the federal government may now claim that specific but undetermined amounts of water either must be released from a reservoir or not be impounded in order to protect the habitat of an endangered or threatened species. These regulatory water rights are different from both federal-reserved proprietary rights and state appropriated rights. The ESA's regulatory water rights are more open-ended than reserved rights in that they have no priority date and do not depend on the express or implied intent of Congress. The closest the act comes to recognizing any property rights is in Section 5, which allows the federal government to acquire land and other resources to protect a species and its habitat. Lack of congressional intent, however, is irrelevant. Even though water rights are not expressly claimed in the act, they are the necessary consequence of protecting a species or its habitat. The rights flow from the structure of the act—as such, it is easy to understand why the act is viewed as disruptive in the West. (See, generally, Tarlock, 1985.)

Two scenarios are possible. The first is that regulatory water rights claimed under the act will be perceived as fundamentally inconsistent with western water law. In this case there will be increasingly bitter conflicts between those seeking to divert water for agricultural, municipal, or industrial use and those seeking to protect endangered fish and wildlife. Because regulatory water rights are federal and most consumptive water rights are state-created, the conflict will add another chapter to the long and acrimonious history of federal/state water conflicts, especially in the West. In the second scenario, regulatory water rights and state water law will become more congruent as state law shifts to the greater protection of nonconsumptive uses. As we shall see, both scenarios are possible.

The first scenario reflects the historic tradition that western water rights are consumptive rights. Western states follow the law of prior appropriation (subject to dual riparian rights in California, Nebraska, Texas, and Washington). Classic prior appropriation is premised on six basic assumptions. First, water is owned by the state in trust for the public; hence, the acquisition of private rights can be regulated and all water rights are usufructuary. Second, the optimal use of water will be served by a system that maximizes private use of water and minimizes public uses for such purposes as instream flow maintenance. Third, private rights should be as secure as possible subject to the well-established principle that claims cannot be asserted for speculative purposes. Fourth, rights are based on the priority of application to a beneficial use (subject to relation back to the date of filing) and endure as long as the claimant applies the water to a beneficial use and does not abandon the use or suffer a forfeiture. Fifth, the whole stream may be diverted during times of peak demand to satisfy calls on it. And sixth, a call may be rejected by the stream administrator only if it would be futile.

Regulatory water rights, on the other hand, more closely resemble public rights. The second scenario is premised on the greater recognition of these rights in western water law. Historically, public rights specify the types of permitted and prohibited nonconsumptive uses that can take place with respect to a body of water. Public rights, however, do not directly give the right holders (the government and its citizens) an entitlement to a specific quantity of water. For example, the public has long had the right to make use of navigable water for commercial and recreational purposes. Public proprietary rights also exist in the West; federal-reserved Indian and non-Indian rights are the most important of these.

Public water rights generally have been accepted because of their long history, but western water lawyers have been troubled by federal-reserved public proprietary water rights for both theoretical and practical reasons. Theoretically, federal-reserved proprietary water rights are mules. They have hybrid appropriative and riparian characteristics. They have a priority date and entitle the holder to a fixed quantity of water just like appropriative rights. But they also have riparian characteristics in that such rights depend on federal landownership rather than on the application of water to a beneficial use. These characteristics have given rise to the concern that huge amounts of water will be claimed by their beneficiaries: Native American tribes and federal land management agencies. The fear of western states is that the exercise of these "phantom rights" will destabilize the intricate fabric of western water allocation. To date, these fears have proved groundless. Not only do reserved rights mimic private water rights and thus have internal, self-limiting characteristics, but there is judicial and congressional hostility to their recognition.

Public rights have historically played a marginal role in western water law, but that role is changing dramatically. As a result, western water law is becoming more congruent with the goals of the Endangered Species Act. What seems to be emerging out of recent water adjudications is that state-created water rights are not different from any other property rights despite the vast energy exerted by western water lawyers to urge a contrary result. Thus, state water rights are not immune from the retroactive application of state police power or federal constitutional authority. Consumptive water rights may be trimmed to protect public rights to navigation and environmental quality. Although the public has long had the right to make use of navigable water for commercial and recreational purposes, this right was seldom in conflict with consumptive uses. The public right of navigation and the law of consumptive rights are no longer parallel doctrines, however, as states expand the public trust doctrine and sharpen the conflict between consumptive and nonconsumptive uses.

California, for example, has used the public trust doctrine to integrate public and private rights. In 1983, the California Supreme Court joined the public rights of navigation with classic appropriation law by holding that all vested appropriative rights are subject to the public trust.[2] The public trust, as previously construed by the California Supreme Court, encompasses the protection of fish and wildlife. Building on *National Audubon*, an intermediate appellate court has held that water quality planning must be done independently of any consideration of vested water rights. Specifically, it held that the California State Water Resource Board has to establish minimum flow schedules to protect the salinity balance in the Sacramento–San Joaquin Delta.[3] To implement the decision, the board currently is considering various conservation strategies to preserve the necessary flows. Other states recently have recognized the preservation of instream flows as a beneficial use. This permits them to perfect instream appropriations and to reserve waters for this purpose. Moreover, these states are beginning to incorporate public trust concepts into water allocation law. Taken together, these developments are creating a new riparianism throughout the West, the net result of which may be to make any new diversion or transfer presumptively invalid. At minimum, instream use protection will be an integral part of new diversions and transfers of existing uses and it will be easier to fulfill ESA duties under state law.

MAJOR CASES

Efforts to weaken the Endangered Species Act have failed—and are likely to fail in the future in light of increasing appreciation of biological diversity. Since the ESA first passed, there has been a fundamental shift in values. Although we cannot see the angle of repose from these changes, enough of a transition in popular thinking has occurred to simply recognize it. To date, regulatory water rights under the act have been created by a series of important federal decisions that reflect a new set of environmental values. Although the cases raise as many questions as they answer, they make it clear that endangered species protection must be added to the list of environmental constraints that affect western water development. The issue is no longer whether water rights for endangered species exist, but under what circumstances and in what manner they can be asserted.

The Grayrocks Dam

The first major clash between the Endangered Species Act and water allocation arose when Nebraska discovered that downstream irrigators on the Platte River could be better protected under the wing of the endangered whooping crane than by litigating the allocation of the river under interstate compacts and the doctrine of equitable apportionment.

In the Grayrocks litigation, the court set aside a federal loan guarantee from the Rural Electrification Administration (REA) and a Clean Water Act Section 404 permit[4] because of the effect of the diversion on the downstream habitat of the whooping crane.[5] Ruling that the REA had failed to consult the Fish and Wildlife Service, the court concluded that the REA finding that there would be no adverse effect on whooping crane habitat was insufficient without the Fish and Wildlife Service's biological assessment. Moreover, the Section 404 permit was issued before the Fish and Wildlife Service had completed its assessment. However, the federal government had reserved the right to impose operating conditions on the reservoir until the Fish and Wildlife study was complete. The court held that to allow reservoir operation before adequate biological information was available did not sufficiently ensure that the whooping crane's habitat would be preserved. A settlement favorable to the whooping crane—and, incidentally, Nebraska irrigators—ended the litigation.

This suit was only one of many recent legal actions to protect the flow of the Platte River. It is an ongoing story that illustrates the effect of species protection on state water law. The National Wildlife Federation has been remarkably successful in preventing the diversion of water out of the Platte Basin, using a mix of Nebraska statutes and common law. In 1980, the state supreme court overruled a long-standing and much criticized decision prohibiting any transbasin diversions,[6] and the legislature then subjected all transfers to a wide-ranging public interest review. The court then ruled that this statute requires that the state water agency consult with the Fish and Wildlife Service when the diversion threatens endangered species.[7]

Transbasin diversions remain a major issue in Nebraska. A recent case raised the issues of whether the State Department of Water Resources had correctly determined that a diversion would not harm the cranes and whether an application could be denied to protect them if no appropriated water were available. But in late 1988, the court ducked these issues and blocked an interbasin transfer on the technical ground that water rights applicants have no transferrable interest. Nebraska has also filed an original action in the U.S. Supreme Court to protect its earlier decreed rights to the Platte River. The case will be the first equitable apportionment action in which environmental issues play a major role.

Riverside Irrigation District v. Andrews

Riverside Irrigation District v. Andrews[8] is the key ESA case to date because it establishes the federal government's power to regulate the possible downstream environmental impact of wa-

ter impoundment and diversion projects under Section 404 of
the Clean Water Act and the Endangered Species Act. In *Riverside*, the proposed dam site was on a tributary of the South
Platte River in Colorado. Because the South Platte and the
North Platte rivers are overappropriated, the Fish and Wildlife
Service was concerned that any additional withdrawals might
jeopardize the habitat of the endangered whooping cranes more
than 200 miles downstream. Although the factual issues are very
complex, the Fish and Wildlife Service's first biological assessment concluded that reservoir releases or diversions were necessary during the spring and early summer to scrub out vegetation
in the river channel. The vegetation enables predators to hide
and prey on the cranes. A subsequent assessment, however, suggested that any flows from Wildcat Reservoir would not provide
adequate scouring flows and that mechanical clearing in Nebraska was necessary.

Initially, the U.S. Army Corps of Engineers refused to issue a
nationwide permit for the discharge of sand and gravel during
construction of the dam. This was clearly the only correct decision, for this was not routine activity with minimal and predictable consequences. The corps' decision was based on the adverse
environmental impact to the cranes' critical habitat from the
operation of the dam, not on the impact of the deposit of dredge
and fill material during construction. The U.S. Court of Appeals
for the Tenth Circuit upheld the corps' refusal to issue a nationwide permit but remanded the case to the district court to determine whether the Corps of Engineers had exceeded its statutory
authority.[9]

On remand, the district court ruled that the Corps of Engineers had not exceeded its authority. The opinion squarely rejected the district's contention that the federal government
lacks the authority to create water rights beyond federal-
reserved proprietary water rights and public water rights such
as navigation servitude.[10] The Tenth Circuit affirmed the district court's decision, but it did so on narrow grounds so that the
broader question about how the ESA fits within existing state
allocation law remains open. The important holding, however,
is that the U.S. Army Corps of Engineers did not exceed its
authority when it denied the district a nationwide permit for the
deposit of dredge material, on the basis of adverse downstream

environmental impact, to construct Wildcat Dam and reservoir.[11] The court rejected the argument that the corps is limited to consideration of either direct, on-site effects of a discharge or changes in downstream water quantity that are the direct effect of a discharge: Section 404 "focuses not merely on water quality, but rather on all of the effects on the 'aquatic environment' caused by replacing water with fill material."

The court also addressed two arguments, based on the Wallop amendment and an interstate compact, that western water rights administration precludes the protection of endangered species. In 1977, western states secured an amendment to the Clean Water Act (known as the Wallop amendment) to try to reduce the possibility that effluent limitations would eliminate return flows upon which junior appropriates depend.[12] The amendment ambiguously declares that it is the policy of Congress that the Clean Water Act should not supersede state water allocations. Yet the court held that the Wallop amendment did not bar the exercise of Section 404 discretion because a "fair reading of the statute as a whole makes clear that, where both the state's interest in allocating water and the federal government's interest in protecting the environment are implicated, Congress intended an accommodation . . . in the individual permit process." Riverside's more far-reaching argument that the South Platte River Compact precluded the denial of a permit was dismissed because the issue was not ripe. Congress retains the power to modify a compact.

Carson-Truckee Water Conservancy District v. Watt

Application of the Endangered Species Act to existing projects may pose the hardest integration questions. To date, however, the courts have held that the act applies equally to existing projects. *Carson-Truckee Water Conservancy District v. Watt*[13] is the leading case that illustrates the problem of modifications to ongoing project operations. In *Carson-Truckee*, the court held that the interior secretary has a duty to operate a reservoir located in California upstream from Pyramid Lake, located in the Pyramid Lake Indian Reservation, to protect endangered and threatened species.[14] All parties agreed that the secretary had a duty to prefer fish to municipal and industrial uses in the allocation of water over and above that involved in a previous adjudication as part of the operation of the Stampede Reservoir. Thus the key issue was to what degree fish should be preferred under the Endangered Species Act. The court held that the secretary had a duty to defer all other uses until the fish were no longer classified as threatened or endangered. In doing so, the court rejected the irrigation district's argument that the act required the secretary to avoid only those actions that jeopardized the bare survival of the species. As a consequence, Carson-Truckee's proposed operating plan for the reservoir was found to be inconsistent with the secretary's species restoration duties under the Endangered Species Act and his fiduciary obligation to the tribe: "Water releases for the fishery in a single year may require all of the Stampede storage, leaving no reserve for [municipal and industrial] users in drought years."

The U.S. Court of Appeals for the Ninth Circuit affirmed the trial court's analysis of the Endangered Species Act.[15] Judge Pregerson reasoned that Section 3 of the ESA defined the term "conserve" as "the methods and procedures which are necessary to bring any endangered species or threatened species to the point at which the measures of the Act are no longer necessary." Thus "the ESA supports the Secretary's decision to give priority to the fish until such time as they no longer need ESA's protection." The court did reverse the trial judge's construction of the Washoe Project Act and held that the secretary was not restricted to water sales for municipal and industrial purposes. This construction enabled the court to avoid the hard question of how the reservoir must be operated in the future. A comprehensive bill to settle most water disputes in the Truckee-Carson river basins is being considered by Congress.[16] In the meantime, endangered species protection remains the sole use of the reservoir approved by the Department of Interior. (See Kramer, 1988.)

Other legislation, such as the National Environment Policy Act, the Federal Power Act, and the Northwest Power Act, also promotes regulatory property rights and extends the Endangered Species Act. ESA considerations may be partially folded into the assessments required under these acts, or they may require greater protection of nonendangered fish.[17]

FINDING MIDDLE GROUND

Based on our experience thus far, at least one lesson seems clear. Judicial ground rules for species protection should provide sufficient incentives for all interested parties to strike creative bargains rather than resorting to the courts to solve future protection claims. Indeed, the federal decisions described here have stimulated interest in alternative ways to meet the obligations of the act and still permit water development. Proposed solutions that have emerged from ESA amendments and from negotiations may move the act more in the direction of species rather than habitat protection and thus must be viewed with some caution.

In 1982, Congress provided a mechanism to allow the "taking" of endangered species when the Department of Interior approves a conservation plan proposed by a private developer, provided that the plan does not appreciably reduce the population of the species. A valid conservation plan can reduce the extent of existing habitat, for example, but improve remaining habitat and create a trust fund to maintain and protect the habitat into the future.[18] The Fish and Wildlife Service has in some instances allowed water developments that would jeopardize an endangered species if the developer agreed to pay depletion charges that are dedicated to conservation and recovery measures.

In the early 1980s, the Upper Colorado River Basin states appeared to be stalled by the interior secretary's proposal that ambitious minimal flows be maintained to avoid jeopardy to the area's endangered fish species. In 1987, after three years of negotiations, representatives of the Fish and Wildlife Service, Colorado, Utah, and Wyoming, the Bureau of Reclamation, water developers, and environmentalists reached consensus on a multifaceted recovery plan for the humpbacked chub. The plan includes such measures as acquisition of instream flows within the state water rights system, releases from federal reservoirs, creation of better habitat, curtailing of planting of exotic fish such as bass, construction of passage facilities, artificial propagation, and creation of a fund from federal and state contributions and one-time payments from new water projects to support continued recovery efforts. A committee representing the state and federal agencies is to oversee the plan's implementation (see MacDonnell, 1985). In 1988, Congress authorized $1.2 million to buy active water rights in the Lahontan Valley to increase freshwater flows into the Stillwater Wildlife Management Area east of Reno, Nevada.[19] Various interested groups are trying to negotiate the maze of state water law and federal reclamation law to develop an effective acquisition strategy.

These efforts may be crucial to the long-run success of the Endangered Species Act. If species protection is considered an integral element of state law, ESA remedies may take the form of state water rights rather than federal regulatory water rights. Yet such efforts must be carefully monitored by the environmental community to ensure that the broad purposes of the act are fulfilled.

Notes

1. *TVA v. Hill*, 437 U.S. 153 (1978).

2. *National Audubon Society v. Superior Court of Alpine County*, 33 Cal. 3d 419, 189 Cal. Rptr. 346, 658 P. 2d 709 (1983), cert. denied, 464 U.S. 977 (1983).

3. *United States v. State Water Resources Control Board*, 182 Cal. App. 3d 82, 227 Cal. Rptr. 161 (1986).

4. Section 404 recognizes the U.S. Army Corps of Engineers' traditional jurisdiction over dredging and filling of navigable waters, but subjects dredge and fill permits to a public interest review and allows the federal EPA to veto a corps permit for environmental reasons.

5. *Nebraska v. REA*, 12 Environmental Rep. Cas. (BNA) 1156 (D. Neb. 1978), appeal vacated and dismissed, 594 D. 2d 870 (8th Cir. 1979).

6. *Little Blue Natural Resources District v. Lower Platte North Natural Resources District*, 206 Neb. 535, 294 N. W. 2d 598 (1980).

7. *Little Blue Natural Resources District v. North Platte Natural Resources District*, 210 Neb. 862, 317 N. W. 2d 726 (1982).

8. 586 F. Supp. 583 (D. Colo. 1983).

9. *Riverside Irrigation District v. Stipo*, 658 F. 2d 762 (10th Cir. 1981).

10. *Riverside Irrigation District v. Andrews*, 568 F. Supp. 586 (D. Colo. 1983).

11. *Riverside Irrigation District v. Andrews*, 758 F. 2d 508 (10th Cir. 1985).

12. 33 U.S.C. 1251(g).

13. 549 F. Supp. 704 (D. Nev. 1982), aff'd, 741 F. 2d 257 (9th Cir. 1984), cert. denied, 470 U.S. 1083 (1985).

14. This holding is particularly significant because the Supreme Court held in a parallel case that an early adjudication which included no reserved rights could not be reopened to include water for Indian fishery maintenance.

15. *Carson-Truckee Water Conservancy District v. Clark*, 741 F. 2d 257 (9th Cir. 1984).

16. S. 1588.

17. *National Wildlife Federation v. Federal Energy Regulatory Commission*, 801 F. 2d 1505 (9th Cir. 1986).

18. *Friends of Endangered Species, Inc. v. Jantzen*, 760 F. 2d 976 (9th Cir. 1985).

19. P.L. 100–446.

References

Kramer, John. 1988. "Lake Tahoe, the Truckee River, and Pyramid Lake: The Past, Present, and Future of Interstate Water Issues." *Pacific Law Journal* 19(4):1339.

MacDonnell, L. 1985. *The Endangered Species Act and Water Development Within the South Platte Basin*. Colorado Water Resources Research Institute, Research Report 137.

Tarlock, A. Dan. 1985. "The Endangered Species Act and Western Water Rights." *Land and Water Law Review* 20(1):1–30.

INVERTEBRATE CONSERVATION

by

Dennis D. Murphy

Conservation biologists have been enjoined by E. O. Wilson (1987) to focus attention on "the little things that run the world"—attention commensurate with the roles of invertebrates in a healthy natural world, rather than with their physical size. In his plea, Wilson rightly argues that should invertebrates disappear, humankind would soon follow—along with fishes, birds, other mammals, and flowering plants. Insects, in particular, in their "staggering abundance," are fundamental to the ability of the earth's ecosystems to deliver countless resources and services. The sheer number of insects and other invertebrates, however, makes their conservation a special challenge to the implementation of the Endangered Species Act (ESA). How can we dream of protecting even an infinitesimal fraction of a group that likely exceeds 10 million species worldwide? How can we rationalize spending precious conservation dollars on members of a group that includes the vast proportion of the most virulent pests of our forests, our crops, and our stored food products? And what should we expect in terms of government protection of our most endangered bugs and their kin?

At the very least, we probably should expect more than we have received to date. Just seventeen insect taxa indigenous to the United States have been conferred federal protection since authorization of the Endangered Species Act in 1973 (USFWS, 1988). To be fair, other invertebrate groups have fared proportionally better: Thirty-one clams indigenous to the United States, nine crustaceans, and eight snail taxa, including one particularly diverse genus, *Achatinella*, currently are listed as federally endangered or threatened. These seventeen protected insect taxa, however, constitute far less than 0.01 percent of the species-level taxa in North America. In contrast, 36 mammal species, subspecies, and populations are listed of some 300 species-level mammal taxa restricted to the United States, and 68 bird species, subspecies, and populations (a large number of which are Hawaiian endemics) are listed of approximately 700 species in the United States. These vertebrate groups "enjoy" representation on the Endangered Species List several orders of magnitude greater than that conferred upon insects.

Certainly a strong case can be made that disproportionate attention paid to vertebrates—especially what have been termed "charismatic megavertebrates" (grizzly bear, gray wolf, bald eagle)—constitutes a most responsible use of endangered species dollars in an era of limited appropriations and burgeoning conservation needs. For the most part, vertebrates elicit the most public sentiment, empathy, and support. Many vertebrate species have experienced extensive and well-documented declines in population sizes and fragmentation of their historical distributions. Moreover, many vertebrates, especially those that occupy relatively high trophic positions within ecosystems or are particularly wide-ranging in their pursuit of resources, often serve as effective "umbrella species" (species whose conservation incidentally confers protection on other species with lesser habitat area requirements).

The obvious importance of vertebrate conservation, however, does not negate the need for increased attention to declining invertebrate species. Public perceptions of invertebrates—combined with the traditional goals of the U.S. Fish and Wildlife Service (FWS) and legal taxonomic requirements for listing invertebrate subspecies—have confounded implementation of the Endangered Species Act. This essay explains in part why invertebrates, and insects in particular, have not received adequate attention from the FWS's Endangered Species Program. It further suggests that effective ecosystem conservation, a recurring theme in this book, may require more assertive listing of threatened and endangered invertebrates.

HISTORICAL MISSION OF THE FWS

The historical mission of the U.S. Fish and Wildlife Service is reflected in agency staff that, not surprisingly, is rich with experience in traditional fish and wildlife issues. Many wildlife training programs treat invertebrates merely as wildlife food, not as organisms with particular intrinsic value. Notwithstanding this tradition, the FWS rose to the challenge of the Endangered Species Act's broad mandate at one point during the 1970s by staffing its Washington office with three specialists in invertebrate taxa. Thirteen insects were listed in the years immediately following authorization of the act (Opler, 1990). These signs of progress, however, were short-lived. Since 1976, the emphasis appears to have shifted back to traditional fish and wildlife values. Today the Washington office has no invertebrate specialists at all, and only four insects have been listed in the last decade.[1]

With recent policy changes, the primary responsibility for domestic listing has been transferred to the seven regional directors. Thus the attitudes of regional office personnel now play a primary role in the listing process. FWS biologists have indicated that top staff in some regions are specifically opposed to future listing of invertebrates. Candidate invertebrate species are increasingly relegated to what is becoming a "Category 1 holding pattern."[2] Although biological evidence justifies immediate listing of a particular species as endangered or threatened, neither staff nor time is available to produce a complete biological opinion or final rule for the *Federal Register*.

The problem of staff bias is acute in the field offices of the FWS, which are, by any measure, grossly underfunded. Certain activities, such as those associated with the listing process for the northern spotted owl in Region One, are so technically demanding and bureaucratically complicated that staff can deal effectively with little more than a single action at a time. At the time of this writing, a substantial portion of the staff of the southern California office is focusing on a habitat conservation plan for the Stephen's kangaroo rat, a species with a distribution unfortunately entirely within the sprawl of suburban Los Angeles Basin. The intensity of this effort has reduced attention to other listed species and candidates, including the Quino checkerspot butterfly, a subspecies that may have recently become extinct north of Baja California.[3]

PUBLIC PERCEPTIONS

The FWS has defended its recent lack of listing actions involving invertebrates by suggesting that because the general public does not support such listing it may undermine support for the Endangered Species Act as a whole. While everyone loves a panda and whales arouse tremendous sympathy (witness the public response to the whales trapped by autumn ice flows off Alaska in 1988), few people appear to appreciate the roles of myriad unsung species that comprise and support the diverse ecosystems upon which such charismatic species depend. Yet it would seem that the lack of ecological understanding would justify an educational campaign rather than the perpetuation of outdated policies. In fact, public support for endangered species protection appears to bridge taxonomic groups.

While few invertebrate taxa are likely to elicit public sympathy as do the megavertebrates, certain invertebrates have more popular appeal than many vertebrates. When Steven Kellert (1986) surveyed American attitudes toward imperiled species, he found that while only 43 percent of respondents would favor modifying an energy project to protect an endangered snake species, fully 64 percent would favor protection of a butterfly species in the same situation. (These responses are compared with extremes of 34 percent for a spider and 89 percent for a bird such as the bald eagle.) These statistics, however, do not necessarily reflect a sense of ethical concern for invertebrates. As Kellert points out, invertebrates are viewed most positively when they are perceived as having particular aesthetic value (butterflies) or utilitarian value (bees). Nevertheless, the range in public attitudes toward various taxonomic groups is substantial and probably represents a miscalculation by those who suggest that focusing attention on conservation of just certain vertebrates reflects the preferences of the public at large.

Whether or not the listing of endangered or threatened species should be based, even in part, on the "popularity" of the species is fodder for extensive argument. But public perception and political attitudes toward the Endangered Species Act have had a substantial impact on the act's implementation. Indeed, public pressure, especially the threat of litigation, has significantly influenced implementation of the act during recent years. That invertebrate conservation has no specific constituency has undoubtedly contributed to the overall neglect of invertebrates by the FWS. Entomological organizations tend to be dominated by those practicing insect control or by amateur collectors, not by those interested in insect conservation. The leading conservation organizations treat invertebrates as an afterthought, if at all, and the only organization specifically devoted to invertebrate conservation, the Xerces Society, is not politically active.

THE PROBLEM OF PRIORITY

The vertebrate bias among U.S. Fish and Wildlife Service staff, as well as the perception that protection of invertebrate species is not publicly supported, is reflected in an informal hierarchy in which mammals are often given priority over birds, birds over cold-blooded vertebrates, and cold-blooded vertebrates over invertebrates—with plants trailing behind them all. Indeed, a hierarchy of this sort was proposed within the FWS but ultimately shelved in the early 1980s. Nevertheless, a theoretical hierarchy is supported by many staff members and continues to affect the decision-making process.

A more serious and insidious policy actually does relegate invertebrates to the bottom of the priority list. The FWS gives priority to species representing monotypic genera over other full species, subspecies, and populations. A disproportionate number of vertebrate species, compared to invertebrate species, fit the first category. This circumstance stems not so much from evolutionary trends and phenetic and genetic differentiation, but from different taxonomic treatments of the two groups. Taxonomic treatment of birds has led to an extreme subdivision of genera that is grossly out of "balance" (Mayr, 1969) with that in nearly all invertebrate groups. For example, the fifteen North American hummingbird species, an especially homogeneous lot, are recognized as representing ten genera. It is difficult to imagine a group of equally closely related insect species represented in more than a single genus (see Ehrlich and Murphy, 1983). This taxonomic priority system produces a substantial bias: More taxa tend to be listed from groups subject to the most egregiously split taxonomies. Conversely, species groups previously subject to appropriately conservative taxonomic revisions, including many invertebrate species groups, suffer the consequences of low conservation priority.

The effects of de facto and formal priority systems are most significant in regions that have already incurred significant losses of biological diversity. For instance, wide-ranging species such as the grizzly bear, tule elk, and pronghorn antelope once thrived in the San Francisco Bay area. These and many other large-bodied, high-trophic-level species were virtually extirpated before California became a state (Murphy, 1987a). In regions now undergoing intensive development, vertebrate taxa that might have served as protective umbrella species are long gone. A priority system weighted toward large vertebrates thus is inherently skewed against endangered species conservation in areas with long histories of human settlement, high densities of human population, and the greatest conservation needs. A paucity of high-priority taxa in such areas constitutes a gap in the conservation safety net that jeopardizes remaining unique habitat types, plant communities, and species assemblages in many urban areas.[4]

Some of the most gravely threatened ecosystems in central California, such as native grasslands, provide an example. Once widespread throughout the state, native grasslands have been replaced in all but a few places by Old World grasses and forbs. Remaining native grassland habitat patches, which support exceedingly rich native plant and invertebrate communities, do not happen to support similarly geographically restricted vertebrate species. The only federally protected animal species that co-occurs with the native grassland plant community is the threatened Bay checkerspot butterfly. This butterfly has a restricted distribution that is only weakly correlated with overall grassland plant species richness (Murphy and Ehrlich, 1989). The Bay checkerspot butterfly currently occupies (and hence confers some protection on) just a dozen native grassland habitat patches out of several hundred such patches in the San Francisco Bay area.

Significantly more grassland invertebrate taxa would have to be listed to provide protection for the full array of plant species scattered across the highly fragmented native grassland landscape.

Given historical rates of invertebrate listings, the ability of invertebrates to confer protection on ecosystems at risk seems small. Under present circumstances, invertebrates confer protection upon co-occurring species only in exceedingly circumscribed remnant habitats (the El Segundo blue butterfly on the remaining Los Angeles coastal dunes, for example) or in ecological communities in which invertebrates hold comparatively high trophic positions (for instance, the recently listed Shasta crayfish).

INVERTEBRATE SUBSPECIES

The authors of the Endangered Species Act noted the necessity of protecting within-species variation by allowing the U.S. Fish and Wildlife Service to confer endangered species protection to entities below the level of full species. That prescient decision, however, was compromised in the first reauthorization of the act in 1978. While protection under the act can be extended to any vertebrate species, subspecies, or distinct population, only formally described species or subspecies of invertebrates can now be listed. Arguments involving the taxonomic status of vertebrate subspecies are rendered moot for purposes of listing. Should disagreements exist over the level of genetic or phenetic differentiation exhibited by groups of vertebrates, they still may be listed as populations. But invertebrates differentiated below the species level must be recognized with a trinomial appellation to be considered for addition to the list. It has long been clear, however, that most subspecies are not discrete entities of evolutionary significance, due among many reasons to the discordant nature of character variation (see Wilson and Brown, 1953). Since subspecies are merely "arbitrary geographical subdivisions of species delimited by variation of one, a few, or many characters" (Murphy and Ehrlich, 1984), the assignment of subspecies within a given species depends on characters selected by the systematist. In the eyes of one biologist, a widespread species may be deemed adequately differentiated to justify the assignment of a dozen or more subspecies names to its constituent populations. To another biologist, that same level of within-species differentiation may justify only a few subspecies names. All else being equal, the likelihood of any subspecies in the latter case meeting the criteria for listing as threatened or endangered is significantly lower than that in a finely differentiated treatment.

The arbitrary nature of subspecies assignments has historically led to disagreements among taxonomists. The FWS is understandably reticent to be drawn into such controversies. Since the FWS depends on "the best available scientific evidence" to support listing decisions, essentially unresolvable taxonomic disagreements do little but hinder the protection process. Such a circumstance led the FWS to allow a proposal to list the Callippe silverspot butterfly (*Speyeria callippe callippe*) to lapse (Murphy, 1990). Traditional treatments applied that name to three populations (on San Bruno Mountain, in the Oakland Hills, and near Vallejo, all in the San Francisco Bay area), making the Callippe silverspot one of the most highly restricted butterfly taxa in North America. The pending publication of a paper (Arnold, 1985) that purported to provide evidence that *S. c. callippe* should include nine previously recognized subspecies (ranging from Mexico to Oregon and east across the Sierra Nevada) squelched the listing. Despite the lack of support from *Speyeria* specialists for the Arnold treatment (Hammond, 1985; Grey, 1989), the FWS has held to its original decision and the Callippe silverspot remains one of the most endangered taxa on the invertebrate candidates list (Murphy, 1987b). That this sort of taxonomic argument can block or substantially delay listings has not been lost on opponents of other specific invertebrate subspecies proposals, such as that of the Mono Lake brine shrimp and the Bay checkerspot butterfly mentioned above (see Murphy, 1988, 1990; Murphy and Weiss, 1988).

RECOVERY PLANS

Few would argue against the value of restoring a threatened or endangered species as a viable self-sustaining member of its ecosystem. The recovery process mandated by Section 4(f) of the

Endangered Species Act justifies, delineates, and schedules recovery actions (USFWS, 1985). Strategies include transplantation and reintroduction of individuals, captive propagation, control of competing species, protection of endangered species from taking, and a range of other activities. And, like most actions carried out under the Endangered Species Act, recovery programs have largely focused on vertebrates. Although vertebrate recovery efforts have provided some notable successes where overexploitation or pesticides constituted the major threat—as in the case of the American alligator and the peregrine falcon—implementation problems have beset many efforts. It would be fair to say that the jury remains out on the overall success of the program.

Despite statements during the early years of the Reagan administration that recovery of species already listed would be emphasized over new proposals and new listings, expenditures on recovery activities have not increased appreciably. Furthermore, as critics of FWS species recovery policy have pointed out, a campaign for devising recovery plans is one thing but real recovery of species is another. While the percentage of listed invertebrates, and insects in particular, with published recovery plans is higher than for threatened and endangered species as a whole (USFWS, 1988), recovery actions to date have been limited. This circumstance reflects not only the priority problems discussed earlier, but also a key problem with recovery planning in general. For small populations of threatened and endangered vertebrates, the appearance of progress toward recovery is most possible. Small populations of large vertebrates are especially subject to further reduction from both habitat loss and the deleterious genetic and demographic effects of small population size. Although programs such as captive breeding often do little to enhance the long-term prospects for small populations, they may at least temporarily alleviate genetic and demographic threats. That opportunity is seldom available for invertebrates. When an invertebrate population becomes so small as to be at risk, from inbreeding or demographic stochasicity, the population will tend either to go extinct rapidly or to rebound in numbers rapidly if the threats to its existence are reduced or reversed (Murphy and others, 1990). In contrast to many vertebrate populations, most invertebrate populations are threatened solely by environmental factors that degrade or destroy habitat—factors that are often difficult to address in a recovery program.

Thus the prescriptions for invertebrate recovery are usually straightforward—acquire critical habitat and manage it to maximize its carrying capacity for the species. Unfortunately, the high price of land around urban and suburban development zones, where many threatened and endangered invertebrates reside, is often prohibitive. Recovery plans that target habitat acquisition may be impossible to implement in cases where appropriations will never be adequate to purchase necessary habitat.

Where land *can* be acquired for invertebrate conservation, FWS management should strive to fulfill the goals of the Endangered Species Act: to provide a means of conserving the ecosystems upon which threatened and endangered species depend. But the great habitat specificity of many invertebrates can frustrate the twin goals of species and ecosystem conservation. Activities that increase local numbers of an endangered invertebrate species may succeed at cost to the more broadly defined ecosystem upon which the species depends. Outplanting of the larval host plants of the endangered Lange's metalmark in its lone remaining dune habitat, for example, may have resulted in significant increases in the population size of the butterfly (J. A. Powell, pers. com.). Some observers, however, indicate that portions of the habitat have become a virtual host-plant monoculture—a circumstance that will undoubtedly harm the conservation prospects for other narrowly distributed species of animals and plants that co-occur with Lange's metalmark, including the federally protected Contra Costa wallflower, the Antioch Dunes evening-primrose, and a wide array of other narrowly distributed but as yet unprotected dune invertebrates.

CRITICAL HABITAT

The bureaucratic definition of critical habitat has varied through the years. As Michael Bean (1983) has observed, the formal critical habitat designation is essentially redundant with language in Section 7 of the Endangered Species Act, which requires that actions must not jeopardize the continued exis-

tence of a threatened or endangered species. That observation—
and the 1982 amendment to the act requiring that economic
assessments accompany critical habitat designations—have
greatly reduced those designations. Nonetheless, all U.S. Fish
and Wildlife Service jeopardy opinions must be based on the
best available scientific evidence concerning the potential impact of activities on habitats crucial to the persistence of a listed
species.

Jeopardy interpretations involving invertebrates can be particularly complex. Whereas habitats temporarily unoccupied by
migrating songbirds or anadromous fishes are generally understood to be essential to the persistence of these organisms, the
importance of temporarily unoccupied but suitable habitat for
invertebrates is not universally recognized (Murphy and Rehm,
1990). For invertebrates, especially sedentary, host-specific species (a description that fits most currently listed invertebrates),
the extinction of local populations can be a common occurrence.
Such species persist regionally as metapopulations (regional
collections of interdependent local populations) supported by
mosaics of habitat patches. Each habitat patch has a certain
likelihood of supporting a population at a given time. This is in
part determined by the size of the habitat patch, its topographic
and vegetational diversity, and its distance from sources of colo-
nists. The stability and persistence of a metapopulation can be
highly dependent on the location and condition of habitat
patches that may be only infrequently occupied or may serve
as "stepping stone" habitats during recolonization periods following local extinction (Gilpin, 1987; Harrison and others,
1988).

At least some ecologists in the FWS recognize the importance
of metapopulations in conservation. Conversion of that understanding into clear and emphatic policy, however, is another
thing altogether. Not unexpectedly, the service has been reticent
to invoke the full strength of the act's prohibitions against incidental taking of listed invertebrates in habitats on the margins
of distributions or against losses of suitable but temporarily
unoccupied habitat within the distribution of the entity. Under
Sections 7 and 10(a) of the act, the common resolution of invertebrate habitat loss is "mitigation" through purchase of unprotected habitat (usually considered to be superior in quality

or greater in quantity) elsewhere. But mitigation plans of this sort result in net losses of threatened or endangered species habitat.

FUTURE PROSPECTS

The preceding list of drawbacks to the utility of the Endangered Species Act as a tool for invertebrate protection is certainly not exhaustive. It does suggest, however, that the efficacy of the act would be significantly greater if some simple changes in implementation were made. For instance, the FWS needs to produce truly practical invertebrate recovery plans and to recognize and fully protect suitable habitat (occupied or not) for threatened and endangered invertebrates. The more pervasive problems of staff bias and priority schemes that essentially exclude invertebrates from fair representation among protected species will be harder to remedy.

Healthy functioning ecosystems upon which endangered invertebrate species—for that matter all species—depend are best protected when the goals of species conservation and habitat (or ecosystem) conservation are concordant. Where suites of species are at risk, the FWS should target individual species—vertebrate, invertebrate, or plant—that serve particular functions in an ecosystem (Ehrlich and Murphy, 1987): umbrella species; keystone species, whose roles in ecosystems are so important that their loss could precipitate a cascade of extinctions (Gilbert, 1980); and indicator species, which as subjects of monitoring schemes indicate the condition of their habitat or the status of other species in the same habitat (Landres and others, 1988). In many ecosystems, the best candidates in these categories are invertebrates. Wide-ranging bees might be appropriate umbrella species in certain diverse floral communities. Many ant species play significant keystone roles, providing critical ecosystem services such as cycling plant and animal products and defending resources against overexploitation by predators. Aquatic invertebrates serve as highly sensitive indicators of freshwater environmental quality.

If, as these arguments suggest, invertebrates should be afforded increased protection under the Endangered Species Act, a formal change in the listing priority system should be considered. The U.S. Fish and Wildlife Service has recognized that the assignment of staff time and appropriations to listing, delisting, and recovery actions is necessary "to make the most appropriate use of the limited resources available" (USFWS, 1985). With that in mind, the candidates under most immediate threat of extinction, regardless of their taxonomic status, ought to receive the most immediate attention. Since the otherwise good intentions of staff and institutional priorities tend to overlook invertebrates, a quota system may be warranted. Annual attention to as few as 2 or 3 percent of the more than one thousand invertebrate taxa on the list of potential candidates would be a significant improvement over recent rates of invertebrate listings. Much is known biologically about a substantial number of these candidates, a factor that could speed the listing process. Furthermore, many invertebrate candidates are currently more endangered by immediate extinction than are many vertebrates and invertebrates already listed. (See Murphy, 1987b, for example.)

These proposed adjustments to the Endangered Species Act, however, ignore the most pressing problem: The single-species focus of the Endangered Species Act has not been especially successful in protecting functioning ecosystems (Csuti and others, 1987; Hutto and others, 1987; Scott and others, 1987). Heroic and often inordinately expensive efforts to save the vertebrate species at greatest risk have neither halted extinctions in nature nor significantly reduced the number of candidates that require immediate protection. Preservation of the wealth of species on this planet, including many species not yet recognized as threatened or endangered, requires development of a means of identifying the most endangered ecosystems and conferring protection to them along with efforts to conserve individual endangered species. The fact that invertebrate conservation rests on environmental factors (including the extent, diversity, and condition of habitats) and metapopulation structure (see Murphy and others, 1990) suggests that invertebrate conservation is most effective as a product of habitat conservation. Appropriate habitat-based conservation plans for invertebrates may thus serve as an effective bridge between the current Endangered Species Act and the "Endangered Ecosystems Act" that is so urgently needed.[5]

The first step demands a systematic assessment of endangered species candidates in all taxonomic groups. The distributions and habitat requirements of vertebrate, invertebrate, and plant species should be mapped and analyzed in streamlined treatments modeled after the Endangered Species Information System, which catalogs biological information on listed species. Habitats that support the greatest concentration of candidates should be identified, and candidates from those habitats should be proposed for listing. Such a scheme should not be taxonomically biased; and, at the very least, the institutional bias toward vertebrates would be partially redirected from obligatory listing of the most endangered vertebrate species to priority listing of vertebrate candidates that co-occur with the largest numbers of invertebrate or plant candidates (since the latter vertebrates apparently reside in hot spots of potential extinction). The listing of several species from such habitats or ecosystems would be an efficient use of funds and staff time. Moreover, it could provide for continued habitat protection should an endangered vertebrate species decline to numbers so low that it must be moved into captivity—for example, the dusky seaside sparrow or the California condor, whose habitats have declined significantly in extent since the birds' last individuals were moved to captivity.

An alternative approach to effective invertebrate conservation would integrate present activities carried out under the Endangered Species Act with some sort of systematic identification of ecosystems (or ecological or plant communities). Candidate species should preferentially be listed under the act from the most endangered ecosystems or from the most species-rich ecosystems presently lacking protection in parks, refuges, wilderness areas, and research natural areas. Listing priorities would reflect concern for vanishing ecosystems above that for imperiled species in otherwise healthy ecosystems. Invertebrate subspecies from remnant habitats of just a few acres would receive more immediate conservation attention than vertebrate species from extensive habitats that are comparatively better buffered from habitat loss or disturbance. Because the vast majority of invertebrates have comparatively small home ranges and occur in rather high local densities, protection of invertebrate diversity will not necessarily require extensive habitat reserves. In Part IV of this volume, Michael Scott and his colleagues outline a plan that would greatly benefit invertebrate conservation, as well as conservation of other animals and plants.

Finally, it is widely acknowledged that extraordinary efforts to save a very few species are misplaced in a world threatened with imminent loss of a great portion of its millions of species. A national commitment to the ecosystem conservation necessary to thwart this trend will undoubtedly require a congressional mandate that should coordinate the conservation goals of the endangered species program, the National Park Service, the Wildlife Refuges System, the Forest Service, and the Bureau of Land Management. The policy now being developed by the National Park Service to preserve biological diversity might provide a model for a National Ecosystems Act in much the same way that the 1966 Endangered Species Preservation Act and the 1969 Endangered Species Conservation Act served as precursors to today's Endangered Species Act.

However ecosystem conservation is ultimately realized, it is clear that current species-by-species conservation simply is not going to meet the invertebrate conservation challenge before us. That observation—along with recognition of the critical importance of invertebrates in ecosystem function—suggests that the best conservation policy for invertebrates targets not the invertebrates themselves but the ecosystems upon which they depend. Until a systematic strategy of conserving ecosystems is developed and implemented, the present "one species at a time" policy will continue to miss the key threat to invertebrate diversity: the accelerating loss and fragmentation of natural habitats.

Notes

1. Two of these insects were listed as parts of "ecosystem listings," in which a suite of co-occurring species are listed as a package.
2. See the essay by William Reffalt on the endangered species listing process in Part II of this volume.
3. A petition and evidence supporting emergency listing of this Category 2 species had not been acted upon after nine months.
4. See the essay by Michael Scott and colleagues on gap analysis and species conservation in Part IV of this volume.
5. See the essay by Paul and Anne Ehrlich in Part IV of this volume.

References

Arnold, R. A. 1985. "Geographic Variation in Natural Populations of *Speyeria callippe* (Boisduval) (Lepidoptera: Nymphalidae)." *Pan-Pacific Entomology* 61:1–23.

Bean, M. 1983. *The Evolution of National Wildlife Law*. Rev. ed. New York: Praeger.

Csuti, B. A., J. M. Scott, and J. Estes. 1987. "Looking Beyond Species-Oriented Conservation." *Endangered Species Update* 5(2):4.

Ehrlich, P. R., and D. D. Murphy. 1983. "Butterfly Nomenclature, Stability, and the Rule of Obligatory Categories." *Systematic Zoology* 32:451–453.

Ehrlich, P. R., and D. D. Murphy. 1987. "Monitoring Populations on Remnants of Native Habitat." Pp. 201–210 in D. Saunders and others (eds.), *Nature Conservation: The Role of Remnants of Native Vegetation*. Sydney, Australia: Surrey Beatty.

Gilbert, L. E. 1980. "Food Web Organization and the Conservation of Neotropical Diversity." Pp. 11–33 in M. E. Soulé and B. A. Wilcox (eds.), *Conservation Biology: An Evolutionary-Ecological Perspective*. Sunderland, Mass.: Sinauer Associates.

Gilpin, M. E. 1987. "Spatial Structure and Population Vulnerability." Pp. 125–140 in M. E. Soulé (ed.), *Viable Populations for Conservation*. Cambridge: Cambridge University Press.

Grey, L. P. 1989. "Sundry Argynine Concepts Revisited." *Journal of the Lepidoptera Society* 43:1–10.

Hammond, P. C. 1985. "A Rebuttal to the Arnold Classification of *Speyeria callippe* (Nymphalidae) and Defense of the Subspecies Concept." *Journal of Research on the Lepidoptera* 24:197–208.

Harrison, S., D. D. Murphy, and P. R. Ehrlich. 1988. "Distribution of the Bay Checkerspot Butterfly, *Euphydryas editha bayensis*." *American Naturalist* 132:360–382.

Hutto, R. L., S. Reel, and P. B. Landres. 1987. "A Critical Evaluation of the Species Approach to Biological Conservation." *Endangered Species Update* 4(12):1–4.

Kellert, S. 1986. "Social and Perceptual Factors in the Preservation of Animal Species." Pp. 50–73 in B. G. Norton (ed.), *The Preservation of Species: The Value of Biological Diversity*. Princeton, N.J.: Princeton University Press.

Landres, P. B., J. Verner, and J. W. Thomas. 1988. "Ecological Uses of

Vertebrate Indicator Species: A Critique." *Conservation Biology* 2:316–329.

Mayr, E. 1969. *Principles of Systematic Zoology.* New York: McGraw-Hill.

Murphy, D. D. 1987a. "Challenges to Biological Diversity in Urban Areas." Pp. 71–76 in E. O. Wilson (ed.), *Biodiversity.* Washington, D.C.: National Academy of Sciences Press.

Murphy, D. D. 1987b. *A Report on the California Butterflies Listed as Candidates for Endangered Status by the United States Fish and Wildlife Service.* Report C-1755 to the California Department of Fish and Game.

Murphy, D. D. 1988. "Ecology, Politics, and the Bay Checkerspot Butterfly." *Wings* (Spring): 4–8, 15.

Murphy, D. D. 1990. "Invertebrate Subspecies and the Endangered Species Act." *Endangered Species Update*, in press.

Murphy, D. D., and P. R. Ehrlich. 1984. "On Butterfly Taxonomy." *Journal of Research on the Lepidoptera* 23:19–34.

Murphy, D. D., and P. R. Ehrlich. 1989. "Conservation Biology of California's Remnant Native Grasslands." Pp. 201–211 in L. F. Huenneke and H. Mooney (eds.), *Grassland Structure and Function: California Annual Grasslands.* Dordrecht: Kluwer Academic Publishers.

Murphy, D. D., and K. M. Rehm. 1990. "Unoccupied Habitats and Endangered Species Protection." *Endangered Species Update*, in press.

Murphy, D. D., and S. B. Weiss. 1988. "Ecological Studies and the Conservation of the Bay Checkerspot Butterfly, *Euphydryas editha bayensis.*" *Biological Conservation* 46:183–200.

Murphy, D. D., K. E. Freas, and S. B. Weiss. 1990. "An 'Environmental Metapopulation' Approach to Population Viability Analysis for a Threatened Invertebrate." *Conservation Biology* 4:41–51.

Opler, P. A. 1990. "North American Problems and Perspectives." In N. M. Collins and J. A. Thomas (eds.), *The Conservation of Insects and Their Habitats.* London: Royal Entomological Society.

Scott, J. M., B. Csuti, J. D. Jacobi, and J. E. Estes. 1987. "Species Richness: A Geographic Approach to Protection of Future Biological Diversity." *Bioscience* 37:782–788.

U.S. Fish and Wildlife Service (USFWS). 1985. *Endangered and Threatened Species Recovery Planning Guidelines.* Washington, D.C.: U.S. Department of Interior, Fish and Wildlife Service.

U.S. Fish and Wildlife Service (USFWS). 1988. "Box Score of U.S. Listings and Recovery Plans." *Endangered Species Technical Bulletin* 12:8.

Wilson, E. O. 1987. "The Little Things That Run the World (the Importance and Conservation of Invertebrates)." *Conservation Biology* 1:344–346.

Wilson, E. O., and W. L. Brown. 1953. "The Subspecies Concept and Its Taxonomic Application." *Systematic Zoology* 2:97–111.

Yaffee, S. L. 1982. *Prohibitive Policy: Implementing the Federal Endangered Species Act.* Cambridge: MIT Press.

PREDATOR CONSERVATION

by

KEVIN BIXBY

WITH REGARD TO PREDATORS,[1] the Endangered Species Act
(ESA) is a very recent attempt to reverse the momentum of
history. Large predators such as wolves and brown bears were
long ago cleared from most of Europe. The same process has
been at work in North America, where predators have been
subject to more or less continuous persecution for the past five
centuries.[2] Motives and methods varied, but the end result was
the same: The larger predators gradually disappeared from settled regions.

The historical treatment of predators, at least in the United
States, is noteworthy on two counts. The first is the zeal with
which their eradication has been pursued. Until quite recently,
predators were widely regarded as "bad" animals and the desir-
ability of killing them went unquestioned. This blanket of social
approbation permitted a campaign of violence against predators that seems appalling in its scale and viciousness today.
Secondly, the decline of predators has generally been a one-way
proposition. Unlike game species, predators have seldom enjoyed the patronage of politically powerful constituencies; consequently, there has been little impetus to restore depleted
populations. While early conservationists such as Theodore
Roosevelt and Aldo Leopold were championing the cause of
game management and restoration, they were at the same time
leading efforts to eradicate predators.[3] The result is that, by and
large, predators remain more widely depleted than other types
of wildlife.[4]

The conservation of predators is complicated by ecological and economic factors. Because of their position at the upper level of the food chain, predators are inherently rarer than herbivores and require larger areas for their conservation. This is clearly a disadvantage in a world with increasingly less room to spare for wildlife. To make matters worse, predators tend to compete with humans for prey. Large predators such as wolves, coyotes, cougars, and bears occasionally kill domestic livestock, and they regularly eat wild ungulates that otherwise might be taken by human hunters. Competition for other prey is less obvious but equally problematic. Commercial shell fishermen, for example, harbor little affection for sea otters since a population of otters can quickly ruin an abalone fishery. Even when a species does not compete directly with humans for prey, its spatial and habitat requirements can be controversial. The cost of maintaining old-growth forest habitat for a pair of spotted owls in the Pacific Northwest, for example, has been estimated at upwards of $600,000 per year in forgone timber harvests.[5]

Another complication is our lingering collective ambivalence toward predators. Although positive views toward predators seem to be in ascendance, old attitudes die hard. Predators are intensely disliked in some quarters and viewed with mild apprehension by most people. Both the federal government and the states are still involved in the business of killing predators as part of animal damage control programs, underscoring the fragile nature of our truce with listed predators. Under the best of circumstances, public support for endangered species stands in roughly inverse proportion to the economic hardship caused by their conservation. Given that the negative economic impact of predators tends to be more widely felt relative to other types of wildlife, the commitment to protecting them under the ESA cannot be taken for granted. Groups that are affected adversely, such as ranchers, often comprise formidable political lobbies and have been able to muster opposition to predator recovery programs at the highest levels.

THE RECORD

Given the challenges, it is remarkable that any progress in recovering listed predators has occurred at all. Yet progress has been made. Although there are fewer red wolves, black-footed ferrets, California condors, and Mexican wolves living in the wild today than when the act was passed, there are more of each in captivity—a circumstance arguably responsible for their survival at all. Outside zoos, gray wolves are returning to their former haunts around the Great Lakes and in the northern Rockies. After a long troublesome period of population stability, sea otters are once more on the increase in California and work has begun on establishing a second population (Carl Benz, pers. com., 1989). The status of grizzly bears in the northern continental divide region has apparently improved enough to prompt proposals for delisting. Large sums of money have been spent with some success on peregrine falcon restoration. Bald eagles are making a comeback. Alligators have rebounded with a poodle-eating vengeance. And the red wolf has been reintroduced into the wild along the Carolina coast. Recovery officials cannot take all the credit for these achievements, but none would have occurred without protection under the act.

Perhaps the most important accomplishment of the ESA as far as predators are concerned has been to put an end to centuries of indiscriminate killing. The act did not change attitudes toward predators overnight, but it did establish substantial disincentives for previously condoned behavior. Under the act, the taking of a listed species is a federal crime punishable by stiff fines and prison sentences.[6] The threat of prosecution has not eliminated illegal killing, but it has reduced the wanton destruction of predators. This by itself has given endangered predators a little breathing room. Wolves in Minnesota and the northern Rockies, for example, have recolonized former habitat on their own since being protected.

Another major accomplishment has been to stimulate interest in—and funding for—predator research. Prior to the ESA, there was little of either. The ESA has not only made federal and state funds available for research on listed species, it has also elicited private-sector research support. Much of what is known about grizzly bears in the Cabinet-Yaak ecosystem in the northern Rockies, for example, is the result of studies funded by U.S. Borax Corporation.[7]

The act has also managed to promote a certain level of cooperation among agencies with divergent missions. Such coordination is vital to predator recovery since predators tend to wander over large areas and cross multiple jurisdictional boundaries. The most prominent achievement of the grizzly bear recovery program has been the establishment of an Interagency Grizzly Bear Management Committee (IGBC) comprised of the regional directors of the Fish and Wildlife Service and National Park Service, several Forest Service regional foresters, the Montana state director of the Bureau of Land Management, and representatives from the wildlife agencies of Idaho, Montana, Wyoming, and Washington. Although the IGBC has been criticized by some observers for being more protective of agency turf than of bears, it has at least created an administrative structure in which grizzly bear recovery is regularly discussed.

Finally, the role of the ESA in changing public attitudes toward predators should not be overlooked. There is no question that a dramatic reversal of anti-predator attitudes has taken place over the past twenty years (Kellert, 1985). This shift cannot be attributed entirely to passage of the ESA, of course. But insofar as any law hastens public acceptance of nascent ethical tenets, the act has helped to promote the belief that predators, along with other species, deserve protection.

Despite these successes, the overall record of predator recovery under the act is disappointing. Intractable political opposition, local public resistance, and a hiatus of high-level administrative support for endangered species conservation during the Reagan years have prevented managers from taking the bold measures needed to bring about recovery for most listed predators. With two notable exceptions (the sea otter and the red wolf), tangible progress toward delisting is evident only in cases where conditions have favored natural recovery and managers have not been required to do anything very controversial.

ISSUES AND TRENDS

New Phase

It appears that predator recovery is entering a new phase. The "easy" part is over. Taxonomic questions have been resolved, recovery plans have been formulated, emergency evacuations from the wild have been carried out, and captive breeding programs have been established. The ironic legacy of past eradication efforts is that these steps will not be enough to achieve recovery objectives. Entirely new populations of predators must be established, and existing populations must be bolstered. Whether this means releasing captive-raised animals into the wild or transplanting wild stock, the next stage promises to be enormously controversial. Even under the best of circumstances it is difficult to gain public support for reintroducing a listed species, since the protection of the act imposes real and imagined constraints on human activities. When predators are involved, the problem is compounded.

With few exceptions, the record of successful reintroductions for listed predators is dismal. All attempts to reintroduce gray wolves, for example, have failed. When the Fish and Wildlife Service asked the three states within the historic range of the Mexican wolf to recommend reintroduction sites, the Texas legislature responded by making it illegal to reintroduce wolves in the state (Brown, 1988). The only site suggested—the remote and inhospitable White Sands Missile Range in southern New Mexico—was removed from further consideration after the army rescinded its earlier approval, ostensibly for security reasons but more plausibly due to pressure from local ranchers.[8] The Fish and Wildlife Service's policy now is to wait until a cooperative land management agency can be found before going forward with reintroductions (Thomas Smylie, pers. com., 1989).

Similarly, the proposal to reintroduce wolves into Yellowstone has been thwarted by ranchers, hunters, and the Wyoming congressional delegation. Just when the prospects for reintroduction looked brightest in 1987, the Wyoming delegation flexed its political muscle and the Fish and Wildlife Service and National Park Service quickly backed away from earlier commitments to wolf reintroduction. In the case of the eastern timber wolf, a reintroduction attempt was at least made, but with little prior effort to garner local public support. Of four wolves released in Michigan's upper peninsula in 1976, all were killed within a matter of months (Hull, 1985). Since then, recovery officials have elected to focus on natural recolonization to achieve recovery goals.

Negotiated Recovery Actions

Experience suggests that biological questions will be increasingly overshadowed by political issues in predator recovery. In the case of Yellowstone, for example, Fish and Wildlife Service officials admit that the obstacles to wolf recovery are purely political at this point. Accordingly, there is growing interest in designing recovery programs to meet political rather than biological demands.

While the Endangered Species Act does not authorize direct negotiation of recovery actions, it does provide recovery officials with considerable flexibility to manage protected species, particularly those listed as threatened. The act authorizes the secretaries of interior and commerce to issue regulations deemed "necessary and advisable" to conserve threatened species.[9]

Congress significantly expanded this latitude when it created the experimental population designation in 1982.[10] Any new population of a listed species established by relocating animals may be designated as an experimental population. Regardless of the status of the species elsewhere, an experimental population is managed as a threatened species or as a candidate for listing, depending on whether it is considered essential to the survival of the species. In either case, restrictions on taking are less stringent than they would be otherwise, making it easier to gain public and agency acceptance for reintroductions.

The red wolf recovery program illustrates how recovery actions can be profitably tailored to win public support. The current effort to repatriate captive-raised red wolves as a nonessential experimental population at Alligator River National Wildlife Refuge in North Carolina is actually the second of its kind. The first, which focused on the Land Between the Lakes region of Tennessee and Kentucky, was stymied by strong opposition from various quarters that forced both states to withdraw their cooperation (Parker, 1990). For the Alligator River project, Fish and Wildlife Service officials made a concerted effort to consult local residents and refuge user groups at an early stage in the recovery planning process. Specific concerns were addressed: Hunters were assured that they could continue using dogs to hunt for deer on parts of the refuge, and trappers were allowed to continue using certain areas. The fears of local residents were assuaged by a promise that all released wolves would be equipped with radio-controlled tranquilizer collars so that errant wolves could be instantly immobilized if needed. Most important, public opinion was solicited before the reintroduction plan took on the appearance of a *fait accompli*. The payoff came when four wolves were released in 1987. A handful more have been released since then, and although more than half have died for a variety of reasons, none has been maliciously killed.

A similar process of de facto negotiation has allowed recovery officials to begin establishing a second sea otter colony at San Nicholas Island off the California coast. The new population is needed to protect the subspecies against extinction in the event of an oil spill or other disaster. Until recently, plans to establish an additional otter colony had been blocked by the oil industry, which feared otters would hinder offshore oil and gas drilling, and by commercial fishermen and abalone divers. Congress addressed the issue during efforts to reauthorize the ESA in 1985. A compromise plan for establishing and managing a new otter population was worked out and enacted as Public Law 99-625. In a major concession to opponents of the new colony, the law specifies that Section 7 of the ESA will apply only to otters in the vicinity of the translocation site. Otters that stray out of this zone are to be retrieved and returned to either the new or parent population (USFWS, 1987).

Yet as promising as these success stories appear, there are several problems with negotiating recovery actions. The first is that negotiation and public education take time. If nothing else, the need to solicit and respond to public opinion delays recovery actions, which in turn may lessen the odds for successful recovery. The Cabinet-Yaak grizzly bear population in the northern Rockies, for example, consists of an estimated fifteen bears. Although biologists doubt the population will survive long without bringing in additional bears (see Maguire, 1986), augmentation plans have been delayed while the Fish and Wildlife Service complies with lengthy public participation requirements.

Similarly, the decision by the Fish and Wildlife Service to postpone Mexican wolf reintroduction until a cooperative land management agency can be found means the subspecies will continue to languish in captivity with an ever greater risk of losing genetic variability, as well as the knowledge and fitness needed to survive in the wild. Biologists now wonder whether the high mortality rate of red wolves released at Alligator River—all raised in captivity—might be attributable to a lack of resistance to diseases encountered in the wild (Rees, 1989).

Another obvious drawback of negotiated solutions is that they tend to address political rather than biological needs and may not always be in the best interest of the species. Not only has the Cabinet-Yaak grizzly bear augmentation plan been delayed, but it has also been watered down in response to local opposition. The number of bears to be transplanted has been reduced, and the plan to place grizzly cubs with foster black bear mothers has been dropped entirely.

Zone Management

A second trend in predator recovery is the increasing reliance on zone management schemes. Typically, these include a core area in which recovery of the species is a high priority, a buffer zone, and a much larger area in which the species is not welcome. Besides sea otters, some type of zone system has been proposed or instituted for grizzly bears and wolves in the Rockies and will undoubtedly be used to manage other predators in the future.

Zone management is a mixed blessing. On the one hand it minimizes conflict by providing a geographic scope to species recovery programs. As such, it can be viewed as a recognition that conserving endangered species is really a land-use problem in which hard decisions must be made about how much space humans are willing to yield to other species. On the other hand, zone management schemes have tended not to be implemented with a great deal of magnanimity. Most have relegated species to the most remote and least economically desirable locations available. The red wolf recovery program, for example, focuses on a swampy national wildlife refuge and adjoining bombing range. The only site proposed for Mexican wolf reintroduction was a desolate desert mountain range on an army missile base. San Nicholas Island was selected for a new sea otter colony because its remote location guarantees fewer conflicts with fishermen and abalone collectors than coastal areas. The site also poses no obstacle to offshore oil and gas drilling since the navy already prohibits these activities in the vicinity of the island. Since a location's remoteness or economic value makes little difference from a biological standpoint, it can be argued that it makes sense to minimize the economic impact of recovery where possible. This may be true. But it can also be argued that protecting species by relegating them to remote sites where few people will ever have an opportunity to enjoy them defeats some of the more important reasons for conservation. The ESA itself cites the "esthetic, ecological, educational, historical, recreational, and scientific value" of endangered wildlife as a rationale for their conservation.[11] Most of these values, however, are utilitarian in nature and their realization depends on human access to wildlife.

Zone management not only designates areas where species recovery is a priority, but also where species will *not* enjoy the protection of the ESA. The practical consequence is to reduce the chances of a species ever becoming reestablished outside of designated recovery zones. Since zone management schemes are usually the outcome of controversy, the tendency is to keep core zones to the minimum size necessary to achieve recovery. The end result is that zone management institutionalizes the gap between the legal definition of recovery (which for predators is usually defined in numbers far below former levels of abundance) and a more generous interpretation of the idea that all species deserve to exist in well-distributed, ecologically meaningful numbers.

Consider the sea otter recovery program. As part of the compromise allowing establishment of a new colony at San Nicholas Island, the species will actually be prevented from recolonizing a major portion of its former range. Most of the southern California coast has been designated a management zone in which otters are not welcome. Otters that wander into this area will be retrieved and returned to either the new or parent population. This trade-off was reached despite biologists' predictions that the main sea otter population off California's central coast would expand on its own into the "no-otters" management zone within twenty years (USFWS, 1987).

Federal Resources

If passage of the ESA erected a barricade of last resort against extinction for many predators, then the availability of federal lands and money has provided the means by which that barricade can be defended. It would not be an exaggeration to say that the recovery of certain predators—grizzly bears, for example—is absolutely dependent on federal resources.

Federal involvement in the management of threatened and endangered species, however, has spawned one of the most contentious issues in predator conservation: the preemption of the traditional authority of states to manage wildlife.[12] With respect to predators, the issue came to a head in 1982 when the Fish and Wildlife Service finally acceded to Minnesota's request to allow sport hunting and trapping of wolves. The state's wolf population is listed as threatened under the ESA. State officials had long argued that such measures would help defuse anti-wolf sentiment and reduce depredation problems. The move, however, was blocked by a U.S. Circuit Court which ruled that the ESA expressly prohibits the public taking of listed species except in extraordinary circumstances where it is the only way to relieve population pressures.[13]

The decision, appealed and upheld in 1985, has had ramifications far beyond Minnesota. Among other things, it leaves the legality of Montana's grizzly bear hunt in doubt. Montana allows a small number of bears to be taken by hunters each year. The Fish and Wildlife Service maintains that the hunt is legal because it is in fact needed to relieve population pressures, although the evidence for this is inconclusive. The Minnesota decision also erected a major stumbling block to wolf recovery throughout the northern Rockies by raising the specter that states would be helpless to control wolves should they become reestablished. Reauthorization of the ESA in the Senate was delayed for three years in part by Rocky Mountain state legislators seeking to overturn the Minnesota decision.

In some cases, disgruntled states have refused to cooperate in predator recovery efforts. Minnesota quickly lost interest in wolf management after its sport trapping season was ruled illegal. Montana has dropped out of the wolf recovery program, as well, and would probably not participate in grizzly bear recovery if its annual grizzly hunt were abolished. States that have taken a hard line against predator recovery (such as Colorado, Wyoming, and Texas) might reconsider if hunting of listed predators were allowed. The states' position is that public hunting is the price that must be paid to win public acceptance of predator recovery efforts. Environmentalists argue that allowing protected predators to be hunted would mean a return to the days when predators were shot on sight. In any case, the act is clear on the matter. Unless the law is changed to allow hunting of listed species, which is unlikely, some states will never be more than reluctant partners in predator recovery.

LESSONS

Given the importance of federal resources to predator recovery, and the attitude of the states, it is clear that the responsibility for predator recovery will continue to fall largely on federal shoulders. Of all the resources the federal government has to offer, none is more valuable than its lands. These are, for the most part, the only places wild enough and large enough to support predator populations. Their continued availability is essential to predator conservation.

Yet the reality is that predator recovery programs are faltering, and opportunities for restoring predators to their former haunts are being lost because too many federal lands are *not* available for predator conservation. Mexican wolves, for instance, continue to languish in captivity despite an abundance of suitable reintroduction sites on federal lands in the Southwest. Grizzly bears could be restored to some of these same sites. There are other examples as well. National forests in the Great Lakes region could probably sustain a reintroduced population of cougars, and there is little doubt that Yellowstone could support wolves.

Part of the problem is that while the ESA is the strongest means we have for predator conservation, it is weak on the issue of restoration. Its goal is to prevent extinctions, not to repair damaged ecosystems. For the most part, the FWS has not been able to bring about the reintroductions needed to achieve even minimal recovery, much less those needed for ecologically meaningful restoration. The act requires federal agencies to "seek to conserve endangered and threatened species" and prohibits them from undertaking, funding, or permitting any action that jeopardizes the continued existence of listed species.[14] The emphasis, however, is on listed species that presently reside on federal lands. Under current regulations, a federal agency does not violate the act if it undertakes or permits activities that reduce the prospects for successful reintroductions. Section 7 permits modification of potential habitat as long as members of a listed species and critical habitat are not harmed.[15] The short-sightedness of these regulations is readily apparent in the case of a species that exists only in captivity, such as the Mexican wolf, or one that occurs in subviable numbers in the wild.

A related difficulty is the continuing deference on the part of Congress toward the tradition of state preeminence in wildlife management on federal lands. Historically, the states have been responsible for managing wildlife while the federal government has provided the habitat. This arrangement dates back to a time when federal authority to manage wildlife or preempt state wildlife law was unclear. (See, for example, Bean, 1983.) Since 1912, however, such authority has been unambiguously affirmed by the courts in a number of cases.[16] Nonetheless, Congress has been loath to challenge the states' traditional role, and this reluctance is reflected in statutory and regulatory language requiring federal land management agencies, particularly the Forest Service and Bureau of Land Management, to consult with the states before undertaking significant actions related to wildlife. The practical result is that states exercise an inordinate veto power over such matters as reintroductions on federal lands.[17]

Predators are probably not essential to human survival. In many instances their ecological function has been usurped (albeit crudely) by two-legged hunters. Yet predators enrich our culture in other ways. They are wilderness incarnate—the living embodiment of the wildness that has shaped our identity as a nation and continues to invigorate us. Their absence from the landscape is readily noticed by the sensitive observer. To those who hold the phenomenon of life in balance as the highest good, the disappearance of predators is cause for profound regret.

It remains to be seen whether the Endangered Species Act and those responsible for its implementation can meet the challenge posed by predators. While some success in predator recovery programs has been achieved, another strategy for endangered predator conservation may be needed—one that unequivocally puts federal lands at the disposal of recovery programs. Most important, we should act with urgency. The record of predator recovery programs to date shows the difficulty of restoring predators to their former haunts. Once these places are gone, however, restoration becomes impossible.

Notes

1. For the purposes of this essay, predators are medium to large-size vertebrates that are substantially carnivorous. Examples of predators protected under the ESA include the gray wolf, red wolf, grizzly bear, eastern cougar, Florida panther, southern sea otter, black-footed ferret, peregrine falcon, bald eagle, spotted owl, and American alligator.

2. Longer than that if human hunters played a role in the extinction of saber-toothed cats, short-faced bears, and dire wolves, as some scientists suspect.

3. Roosevelt once described a cougar as a "big horse-killing cat, the destroyer of the deer, the lord of stealthy murder, ... with a heart both craven and cruel" (quoted in Steinhart, 1989). As president, he ordered predator control in the Grand Canyon National Game Preserve and later led his sons and nephew on predator hunts in the area (Borland, 1975). Leopold wrote editorials in support of predator eradication prior to his ecological awakening (Flader, 1974).

4. For example, predators are disproportionately represented on the federal endangered species list, at least among mammals. While predators comprise about one-third of North American mammalian species, they make up more than half of the listed mammals. Predators have also been more widely extirpated from areas now included within the National Forest System. Of the nineteen species reported by Forest Service biologists as having disappeared from five or more forests, fifteen are predators (Bixby, 1988).

5. This is the estimate given by former Assistant Secretary of Agriculture John Crowell (quoted in *Backpacker* magazine, Dec. 1981).

6. 16 U.S.C. §1540 (1988).

7. Borax was anxious to learn how it could proceed with mining in the area without violating the ESA (Harold Picton, pers. com., 1989).

8. This was the conclusion of several informed persons with whom I spoke.

9. 16 U.S.C. §1533 (1988).

10. 16 U.S.C. §1539(j) (1988).

11. 16 U.S.C. §1531 (1988).

12. See John Ernst's essay in Part II of this book.

13. *Sierra Club v. Clark*, 755 F. 2nd 608.

14. 16 U.S.C. §1536 (1988).

15. 50 CFR 402 (1988).

16. The matter was settled decisively by the Supreme Court's 1976 ruling in *Kleppe v. New Mexico* (426 U.S. 529).

17. Regulations implementing the National Forest Management Act (1976), for example, require Forest Service planners to consult with state wildlife agencies and other federal agencies to coordinate "opportunities for the reintroduction of extirpated species" (36 CFR II 219.19(a)(3)). Forest Service planners generally interpret this as a prohibition on undertaking reintroductions without state approval (Bixby, 1988).

References

Banks, Vic. 1988. "The Red Wolf Gets a Second Chance to Live by Its Wits." *Smithsonian*, March, pp. 100–107.

Bean, Michael. 1983. *The Evolution of National Wildlife Law*. Rev. ed. New York: Praeger.

Bixby, Kevin. 1988. "Extirpated Wildlife and National Forest Planning." Master's thesis, School of Natural Resources, University of Michigan.

Borland, Hal. 1975. *The History of Wildlife in America*. Washington, D.C.: National Wildlife Federation.

Brown, David E. 1988. "Return of the Natives." *Wilderness* (Winter):40–52.

Dunlap, Thomas. 1989. *Saving America's Wildlife*. Princeton, N.J.: Princeton University Press.

Flader, Susan. 1974. *Thinking Like a Mountain: Aldo Leopold and the Evolution of an Ecological Attitude Towards Deer, Wolves, and Forests.* Columbia: University of Missouri Press.

Hull, Christopher, 1985. *The Wolf in Michigan*. Lansing: Michigan Department of Natural Resources.

Kellert, Stephen. 1985. "Public Perceptions of Predators, Particularly the Wolf and Coyote." *Biological Conservation* 31:167–189.

Maguire, Lynn. 1986. *An Analysis of Augmentation Strategies for Grizzly Populations: The Cabinet-Yaak Ecosystem as an Example*. A report to the U.S. Forest Service, March 25, 1986.

Parker, Warren. 1990. "A Historic Perspective of *Canis rufus* and Its Recovery Potential." Red Wolf Management Series Technical Report 3, USFWS. (Published in *Proceedings of the Arizona Wolf Symposium, 90*. Asheville, North Carolina.)

Rees, Michael. 1989. "Red Wolf Recovery Effort Intensifies." *Endangered Species Technical Bulletin* 14(1–2):3.

Steinhart, Peter. 1989. "Taming Our Fear of Predators." *National Wildlife*, Feb./March, pp. 4–12.

U.S. Fish and Wildlife Service (USFWS). 1987. *Federal Register* 52(154):29754–29796.

PESTICIDE REGULATION

by

JIM SERFIS

SINCE THE PASSAGE of the 1973 Endangered Species Act (ESA), several controversies have tested the limits of endangered species protection. Like B-rated suspense movies, these controversies may have different casts and directors, but the plots are so similar one wonders if we have seen them all before. Each involves a vested economic interest on one side and an endangered species on the other. Between them lies the question of how to reconcile our moral and legal convictions regarding other species with the inertia of business as usual. A shrimp fisherman must use special devices in his nets to avoid harming sea turtles; Minnesotans are unable to cull wolves to protect livestock and game species; a small fish called the snail darter blocks construction of a multimillion-dollar dam.

As our commitment to protect endangered species continues to evolve, a new controversy has emerged involving the detrimental effects of pesticide use. Most pesticide uses do not harm endangered species. But those that do can have not only immediate effects but also profound long-term consequences that we are only beginning to recognize. Among the documented victims whose deaths have been attributed directly to pesticide use are gray bats in Missouri, bald eagles in Iowa, California condors in California, and brown pelicans in Puerto Rico (Serfis and others, 1986). Immediate effects are not limited to wildlife, however; several herbicides also have lethal effects on endangered plants.

While direct effects are of great concern, it is the less obvious long-term consequences of pesticide use that pose the more serious threat to endangered species—and the greatest scientific and legal challenges to the conservation community. Contaminated fish ingested by a bald eagle or contaminated insects eaten by an Aleutian Canada goose may slowly accumulate enough pesticide to the point at which they inflict physical harm or impair reproduction (USFWS, 1982). Or a Florida panther, after cleaning its paws and fur of pesticides in which it has inadvertently stepped, may sicken because the chlorinated hydrocarbons cause damage to its liver and kidneys. (USFWS, 1983).

In an effort to solve such problems, pesticide use restrictions were proposed by the Environmental Protection Agency (EPA) for counties in almost every state in May 1987. Not surprisingly, such broad measures have caught the attention of every conceivable player in this endangered species melodrama. While federal agencies scramble to carry out the directives of Congress, environmentalists demand aggressive protection and the agricultural sector bemoans the restrictions as unnecessary and costly.

SECTION 7 RESPONSIBILITIES

Perhaps the most far-reaching provision of the ESA is the requirement of Section 7 that each federal agency must take steps to ensure that its programs will not jeopardize an endangered species or its habitat. Applied to pesticide registrations, Section 7 requires EPA to identify pesticide uses that may harm listed species or their habitats and take precautions to avoid adverse impacts related to these uses in consultation with the U.S. Fish and Wildlife Service (FWS). To date, only 2 percent of the thousands of pesticides reviewed have required consultation with the FWS. Many of those that did, however, involved the jeopardy of an endangered species (Slimak and Dickinson, 1986).

The consultation process supplies advice and information. It is not a device to veto EPA's actions. Although obliged to consult with the Fish and Wildlife Service, the EPA has the final power to determine the appropriate regulatory action to avoid jeopardy to a species, provided it falls within the bounds set by the act itself. If a jeopardy opinion is issued, the FWS suggests alternatives that avoid harming the species. These "prudent and reasonable" alternatives usually restrict or modify the use of the pesticide in the habitat of the species. Measures have included the establishment of no-spraying buffer zones within 100 feet of streams and waterways or limitations on various pesticide uses at certain times of the year. The EPA may use the alternatives suggested by the FWS or devise other alternatives, approved by the FWS, to comply with the Endangered Species Act.

CONSULTATION HISTORY

The paper trail of the consultation process between the EPA and the FWS begins in 1980. While there are records for consultations before 1980, they are few in number. And because they were completed by various branches of the EPA, they were markedly different from later consultations (Serfis and others, 1986). In the first several years, the EPA reviewed pesticides for potential harm they might cause to listed species and consulted with the FWS on a case-by-case basis. By evaluating one pesticide registration at a time, the EPA and the FWS determined that in over sixty-five cases the use of a pesticide could jeopardize one or more endangered species (*Federal Register*, 1989). This approach, however, was cumbersome. Since many registrations were processed each year, the workload of processing new registrations precluded evaluating pesticide uses already registered. As a result, the program was fraught with inconsistencies, both in terms of the protection afforded different species and restrictions placed on similar or even identical uses. The case-by-case approach was also found wanting by the pesticide manufacturers because of market inequalities. Since only new registrations were reviewed, the newer uses were more likely to be restricted than older ones.

To eliminate these inequalities and increase the number of pesticide uses reviewed, in 1982 the EPA adopted a new method referred to as the "cluster approach." Under this new system, pesticides with the same use pattern (pesticides used on grain crops, on forests, as mosquito larvicide, or on rangeland, for example) were considered at the same time (*Federal Register*, 1989). The FWS then prepared a biological opinion that included all the endangered species that might be affected by a particular use pattern. Although the cluster approach seemed to speed up the process initially, it was not without problems. The major crop uses of pesticides were covered under the system, but other minor uses escaped review altogether. It was possible for an unreviewed minor use to be approved with no restrictions while the major use of the same pesticide was restricted. Paradoxically, another problem was the time it took to complete a cluster analysis. Although the cluster approach was designed to facilitate the process, final versions of the cluster packages sometimes took two or three years to complete.

COMPLIANCE PROBLEMS

An independent review of EPA's pesticide program in 1986 revealed that the agency was not complying with the act in approximately one-third of all pesticide cases (Serfis and others, 1986). This review determined that the agency had granted the registration of pesticide products before receiving biological opinions from the FWS and had failed to implement recommendations for restricting harmful pesticides. Jack Moore, assistant administrator for pesticides and toxic substances, did not dispute these findings. In fact, he admitted that the agency had not complied with the requirements of the law (Shabecoff, 1986). More revealing than an admission of compliance problems, however, were the reasons Moore gave for the agency's failure. He suggested that there was a lack of precedent to guide agency policy and that EPA staff were reluctant to act on matters not related to human health. In some cases, EPA staff regarded the protection of endangered species as a low priority compared to agricultural production.

To the degree to which these attitudes are shared by other federal agencies, implementation of Section 7 will remain a stumbling block for adequate protection of endangered species. Some of these lapses in compliance will come to the attention of those responsible for endangered species protection, as did the EPA's failures. But many more will go unnoticed. If it were not for the efforts of the nonprofit organizations and the officials in EPA who started the independent review, EPA's compliance problem might never have seen the light of day.

As the problems were revealed, however, the agency's early

reluctance to restrict the use of pesticides was replaced by an urgency to comply with the directives of Section 7. EPA responded to the independent review and subsequent public attention by developing a plan intended to bring the agency into full compliance with the ESA by 1988. Full compliance meant addressing the restrictions recommended thus far in all of the case-by-case and cluster analyses. The primary thrust of the plan was restriction of pesticide uses in areas where they had been determined harmful to one or more endangered species. Restrictions were to be printed on labels for each product. In addition, users were to obtain information bulletins describing where and how to apply the restrictions. These restrictions were to become effective on February 1, 1988, as stipulated in public notices issued May 1, 1987.

All too predictably, perhaps, the program was to become hopelessly entangled in controversy. The agricultural sector's initial response to the new restrictions was to dig in their heels. Their opposition was spurred by the EPA's admission that the maps to be used for delineating the restrictions and issuing jeopardy opinions might have to be reviewed for accuracy. The EPA grew less and less able to organize and support an effective program and clearly failed to explain the restrictions to users and producers. Consequently, the extent of the program's restrictions was misunderstood and overstated.

In retrospect, much of the controversy might have been avoided if the EPA had approached the problem differently. The Endangered Species Act is often perceived as a prohibitive (and often burdensome) statute that leaves little room for compromise. Yet this need not be the case. In fact, the Section 7 consultation process allows a fair degree of latitude for creating protective strategies that provide room for environmentally responsible development. Resolution of the pesticide conflict may rest on the ability of the parties involved to take advantage of the act's opportunities for creative problem solving and negotiation.

EPA's original approach, however, did not take advantage of these opportunities. The EPA identified pesticides that might harm listed species and then placed restrictions on their use in counties where the affected species lived. Yet the agency did not determine whether the restricted pesticide was actually being used in the range of the species for any of the affected counties. By attempting to limit potential problems in as wide an area as possible, the agency overstated the extent of the restrictions. In doing so, it failed to emphasize that the restrictions usually applied only to portions of the affected counties. From the user's perspective, EPA's program first appeared to be a sweeping ban on the use of all pesticides in many counties. Mark Maslyn of the American Farm Bureau Federation described the program as a "ban [on] approximately two-thirds of all pesticides in all or part of one thousand U.S. counties" (*EPA Journal*, 1988). What Maslyn did not realize, or chose not to concede, was that the restrictions sometimes applied to a use that did not even exist and might apply only to a small portion of each county.

Perceived as another backdoor attempt to stop pesticide use, the plan met formidable opposition. The U.S. Department of Agriculture (USDA) entered the stage and attempted to stop the program, arguing that the restrictions placed an unfair burden on farmers. Some USDA officials were reported to have told EPA staff that farmers might even kill off endangered species so that a restricted pesticide could be used. Fearing high costs and the lack of substitutes, farmers became fervent opponents of endangered species protection. Questions over the program's legality were raised, and at least one group, the Mountain States Legal Foundation, gave notice of intent to sue regarding the program.

In response to the opposition, Congress enacted legislation on December 22, 1987,[1] barring EPA's enforcement of protective measures until September 15, 1988. Congress also set into motion changes in the Endangered Species Act. These changes were later incorporated as an amendment to the act in 1988.[2] They directed the EPA to educate users about requirements and programs to protect endangered species, make public disclosures about these requirements and other restrictions, and complete a report detailing the program and future regulations and procedures needed to comply with the act. As a result of congressional action and the intense lobbying effort launched by the agricultural community, the EPA deferred implementation of the program until 1989, stating that more time was needed to develop an effective strategy. Meanwhile, the agency was to prepare for implementation by conducting a public education campaign, refining scientific data, cooperating with other federal and state agencies, and soliciting public comment on revisions to the program.

The newest version of the program, revealed by EPA in July 1989, uses a two-pronged approach (*Federal Register*, 1989). First, the consultation process will focus on individual species instead of pesticide clusters to determine whether an endangered species may be harmed by the use of a pesticide. Based on a species' status and vulnerability to pesticides, the program will first address those species that are most in need of protection. EPA will identify pesticide uses that occur in the habitats of these priority species and then determine whether uses of the pesticide will harm the species. Priority consideration will continue until all the species are reviewed, and then consultations will occur on a case-by-case basis. Second, EPA will determine the lowest amount of a pesticide that may affect a listed species and initiate consultations only when the application rate of the pesticide exceeds these thresholds. This screening process is meant to ensure that consultations will occur only for species that may be affected by actual pesticide use.

The EPA will also bolster the public participation component of the program by encouraging the involvement of states, counties, the agricultural community, environmental groups, and other concerned parties. The EPA will notify the public before initiating a consultation and will allow an opportunity for comment on biological opinions. Moreover, states will be allowed to develop alternatives to be used in specific localities. The alternative measures will not be enforced until January 1991. The EPA will encourage pesticide users to comply voluntarily with measures in 1989 and 1990. While the new approach is promising, success of the program has always hinged on implementation more than design. Success will come only if the agricultural community accepts the new approach as fair and workable.

PARTIAL FAILURE OR QUALIFIED SUCCESS?

Has the Endangered Species Act been used successfully to protect endangered species from pesticide use? Although much of the plot of this drama has yet to unfold, at least three general observations seem pertinent. First, the implementation of such a large protective program is time consuming. While several years have elapsed since potential problems were identified, time was needed to develop an unprecedented program dealing with many complicated factors. Delays occurred because of uncertainty about how to protect endangered species from pesticides, which in turn led to delays caused by political pressures—the latest pressures resulting in a reappraisal of the program's scientific basis. A great deal of time could have been saved if the EPA had tested the acceptability of the program with the affected parties. Crafting the recommended alternatives to fit the technical, logistical, and political realities of the conflict and then ensuring clear communication of these restrictions might have fostered a greater level of acceptability.

But the pesticide issue begs a more fundamental question: Can such a slow and cumbersome process be expected to protect endangered species? Clearly, substantive protection does not occur until a program is implemented and accepted by the affected parties. Protection must wait as the bureaucratic wheels turn slowly. The reality of endangered species protection is that time is needed to devise solutions that balance effectiveness with practicality. It would be a waste of time to speed implementation of the pesticide program by forcing restrictions on the agriculture sector if enforcement of those restrictions were impossible. In hindsight, EPA's administration of the pesticide program should have been accelerated by testing the political winds and constructing a technically practical program.

Second, while the Endangered Species Act may be criticized for lacking the political muscle to speed implementation of protective measures, its strength is a consultation process that offers avenues to resolve conflicts creatively. Because protection may be achieved in many different ways, there is tremendous (though often untapped) potential for devising solutions that protect species while allowing prudent use of certain pesticides. In effect, the consultation process allows for a negotiated arrangement between the agencies involved (and perhaps, indirectly, pesticide users) to seek the least burdensome solution to an endangered species conflict. To tap this potential, EPA must maintain a flow of information to pesticide users, other federal agencies, and the environmental interest groups. In such a forum, pesticide users could tell the regulatory agencies which alternatives are the most practical and, in turn, the agencies and environmentalists could convey which measures would have to be taken to protect endangered species. Not only would this forum allow for fine tuning of alternatives, it would also allow for the dialogue necessary to make the program politically palatable to all the affected parties.

This has been the strategy used by the Agriculture and Wildlife Coexistence Committee in Cameron County, Texas. This committee, made up of state and federal representatives, agricultural producers, and industry representatives, reviewed pesticide uses in endangered species habitat in their area and recommended steps to avoid problems (*EPA Endangered Species Update*, 1989). Not only will local perspectives help the EPA and the FWS construct workable solutions for the county, but these solutions will be more acceptable since they originated from those most directly affected.

Third, the trend of congressional intervention to delay protection is a serious stumbling block in the conservation of endangered species. In the case of the pesticide controversy, Congress has not only caused substantial delays but also has specified how the program will be run. Congress, however, seldom has the technical expertise to legislate a solution to endangered species conflicts. And while members of Congress may not always understand how pesticides harm endangered species, they do understand that to stay in office they must cater to the needs of their constituents, including pesticide users. Consequently, political expediency is usually more important than worries over the well-being of wildlife and plants that may or may not be harmed by pesticide use.

Congressional intervention might be avoided by devoting more effort initially to organization. No program is exempt from the will of Congress. But fine-tuning a program by an administering agency—before it is fully presented—would ensure greater control over its outcome. In hindsight, it is clear that lingering doubts about where and how endangered species would be affected and the validity of the alternative recommendations should have made the EPA take a second look at the program. In its haste to comply with the act, however, the EPA attempted to implement an untested program that was unacceptable to Congress.

If we have learned anything in the years since the Endangered Species Act was first passed, we have learned that to protect endangered species we must not only be good scientists and bureaucrats, but also good politicians. Ultimately, successful implementation of the act will be measured by our ability to use it to avoid rather than create conflict.

Notes

1. H.J. Res. 395.
2. Endangered Species Act Amendments of 1988, Pub. L. 100-478, 102 Stat. 2304 (16 U.S.C. 2306).

References

EPA Endangered Species Update. 1989. "Cameron and Starr Counties, Texas: New Meaning to Peaceful Coexistence." *EPA Endangered Species Update,* January 1989.

EPA Journal. 1988. "The Controversy Over Pesticides and Endangered Species: Two Points of View." *EPA Journal* 14(3): 26–27.

Federal Register. 1989. "Endangered Species Protection Program: Notice of Proposed Program." *Federal Register* 54(126): 27984–28002.

Inside EPA. 1987. "USDA Opposes EPA Plan to Limit Pesticide Use in Endangered Species Ranges." *Inside EPA,* June 26, p. 15.

Serfis, J., R. Tinney, and R. E. McManus. 1986. *The Environmental Protection Agency's Implementation of the Endangered Species Act with Respect to Pesticide Registration.* Washington, D.C.: Center for Environmental Education.

Shabecoff, P. 1986. "Agency Criticized on Wildlife Role." *New York Times,* Aug. 29, p. D18.

Slimak, M. W., and W. Dickinson. 1986. *Endangered Species and FIFRA: An Implementation Plan.* Internal EPA document.

U.S. Fish and Wildlife Service (USFWS). 1982. *Biological Opinion for Phorate.* FWS/OES EPA-81-10.

U.S. Fish and Wildlife Service (USFWS). 1983. *Biological Opinion for Lindane.* Log 4-1-83-018.

PART IV

CONSERVING BIODIVERSITY

Conservation is sometimes perceived as stopping everything cold, as holding whooping cranes in higher esteem than people. It is up to science to spread the understanding that the choice is not between wild places or people, it is between a rich or an impoverished existence for Man.

—THOMAS LOVEJOY

All that lives beneath Earth's fragile canopy is, in some elemental fashion, related. Is born, moves, feeds, reproduces, dies. Tiger and turtle dove; each tiny flower and homely frog; the running child, father to the man, and, in ways as yet unknown, brother to the salamander. If mankind continues to allow whole species to perish, when does their peril also become ours?

—WORLD WILDLIFE FUND

FROM ENDANGERED SPECIES
TO BIODIVERSITY

by

REED F. NOSS

*American conservation is, I fear, still concerned for the most part
with show pieces. We have not yet learned to think in terms of small
cogs and wheels.*

Aldo Leopold, *A Sand County Almanac*

ONE HALF CENTURY AGO, Aldo Leopold, a premier applied ecologist, advocated a conservation ethic that recognized the interdependence of land, water, and biota. Leopold warned against shortsighted approaches to resource management based on economics, recreation, or single-species biology. Instead, he felt that conservation practice should address the health of the land as a whole and that our role in the biotic community must change from one of conqueror to one of "plain member and citizen" (Leopold, 1949).

How far have we progressed toward Leopold's vision? Today the Endangered Species Act (ESA) and other pieces of American conservation legislation are held up as models for the world to follow. Yet most officially listed species are closer to extinction now than when they originally were listed. Some, like the Palos Verdes blue butterfly and the dusky seaside sparrow, are gone forever. Forty percent of listed U.S. species still have no recovery plan. One thousand species known to qualify for protection are precluded from listing, partially because of budget and staff inadequacies in the U.S. Fish and Wildlife Service. But small budgets are not the only impediment to saving species. Politics, controlled by powerful economic interests, often determine which species are listed and which are not, and which listed species receive attention. High-profile "glamour species," such as the bald eagle and California condor, receive millions of dollars for high-tech recovery efforts while hundreds of lesser-known cogs and wheels (especially plants and invertebrates) silently disappear.

Clearly, the ESA is not accomplishing all that it proposed to do. The logistic and financial nightmare of addressing the biological needs of thousands of endangered species and candidates—not to mention the millions of mostly unnamed species that face extinction in the tropics—suggests that levels of organization *beyond* species may be a more efficient focus of our conservation efforts. Too many species are competing for public attention, and the public knows and cares too little about them. In the following pages, I argue for a hierarchical approach to conservation of biological diversity and urge consideration of ecological processes as well as patterns. Species that play pivotal ecological roles, or those of pragmatic value as indicators or flagships, are important elements in this strategy. To complement a focal species strategy, however, I devote major attention to higher levels of organization—the land as Aldo Leopold so eloquently portrayed it. Furthermore, I follow Leopold in acknowledging that a lot more than better science is required to maintain biodiversity and land health. We need a new ethic, and an ethic put into action.

ARE SPECIES TOO TANGIBLE
FOR THEIR OWN GOOD?

You cannot hug a biogeochemical cycle. You cannot hug a species either, of course. But individual animals, and, for some of us, plants, do have appeal. It is the individual—the "proud" stance of the eagle, the "cute" face of the panda, the "majesty" of the tiger—that attracts most people to wildlife conservation causes. The attraction is emotional, not intellectual. Thus most

Americans may be willing to make some economic sacrifice to protect some endangered mammals, birds, or butterflies, but not to protect most endangered invertebrates, plants, or snakes.

As a collection of individuals, species too can be tangible. But ecosystems composed of nutrient cycles, energy flows, and disturbance regimes in addition to populations are unlikely to capture the public imagination. Only about 7 percent of Americans show a strong "ecologistic" (that is, holistic) attitude toward nature (Kellert, 1980a). Hence we have an Endangered Species Act but not an Endangered Ecosystems Act. After more than a decade and a half under the former measure, we face some difficult questions: How does biodiversity as a whole fare under a species-centered approach to conservation? Can a species-centered approach protect ecosystems? Although the ESA proposes to "provide a means whereby the ecosystems upon which endangered species and threatened species depend may be conserved," habitat protection under the act is incomplete at best (Sidle and Bowman, 1988). The ever-growing list of endangered species, the continual population declines of most listed species, the disruption of ecological processes around the globe—all demonstrate the inadequacies of the ESA at both species and ecosystem levels of organization.

Herein lies the irony: Species collectively may be served best not by a species-by-species approach but by a conservation strategy that transcends individual species. Species after all depend on functional ecosystems. And a focus on collective properties of species (richness, endemism) is an efficient way to identify critical areas for conservation. (See Scott and colleagues in Part IV of this volume.) Some biologists (such as Soulé, 1987) argue that most of conservation history has been concerned with "protection of whole systems" as exemplified by the establishment of national parks; attention to species biology and population viability is more recent. National parks were not set aside to protect systems, however, but rather "to conserve the scenery and the natural and historic objects and the wildlife therein, and to provide for the enjoyment of the same" (National Park Service Organic Act of 1916). Because parks are generally too small for viable populations of many species and their legal boundaries do not conform to ecological boundaries, disruption of processes (such as fire regimes) and species composition is almost inevitable (Kushlan, 1979; Newmark, 1985). Scenery is a hollow virtue when ecological integrity has been lost.

A PLURALISTIC STRATEGY

Biological diversity (biodiversity) is the most popular buzzword in conservation these days, and rightly so. Here, at last, is an opportunity to expand beyond the fragmentary species-by-species approach and address environmental problems holistically. But abandoning species in favor of the ecosystem (as functionally defined) would be an inappropriate response to the biodiversity crisis. Ecological processes continue to function, though perhaps not optimally, even after much of the native biodiversity has been lost from an area. Thus attention to species composition, particular populations of species, and collective patterns of species distribution (floras and faunas) will always be valuable in conservation. But as the catalog of endangered species expands into the millions and threats to the global atmosphere intensify, the species approach is no longer sufficient.

Nature can be understood as a nested hierarchy of wholes. (See Koestler, 1967; Allen and Starr, 1984; O'Neill and others 1986.) Thus our conservation strategy should be pluralistic and address multiple levels of biological organization—from genes to the entire biosphere. Otherwise we might miss something. Recent literature (such as OTA, 1987) has emphasized three primary levels of biodiversity: genetic, species, and ecosystem diversity. Another level being discussed with regard to land-use planning and the management of public lands is the entire landscape or region. No level is fundamental. Which level we focus on depends on the research question or management problem at hand. In this light, the Endangered Species Act should not be treated as a panacea. Rather, it is one element in a broader strategy for conserving biodiversity.

Franklin and colleagues (1981) have argued that structural and functional attributes of ecosystems are just as important as composition. Each of these three primary attributes can be expanded into a nested hierarchy in which specific, measurable elements of biodiversity can be identified at several levels of organization. The genetic level, for example, includes particular rare alleles (composition), the genetic structure of a population, and population genetic processes. The ecosystem level includes species composition and diversity, habitat structure, and ecological processes such as disturbance, decomposition, regeneration, and energy flow. An effective conservation strategy must be broad enough to encompass all these considerations.

Although the identification of measurable "indicators" of biodiversity is an important task, biodiversity is not a purely quantitative phenomenon. At regional and local scales, many important changes in biodiversity are qualitative—for example, the invasion of weedy opportunistic species at the expense of sensitive native species. We trade condors somewhere for coyotes everywhere. Qualitative changes at local and regional scales reflect a homogenization of floras and faunas and yield a net reduction of global biodiversity. When assessing the biodiversity of an area, we need to consider the larger context in which that area exists: What is the regional and global significance of this area, and how can we manage it to contribute most to overall global diversity?

For biodiversity to form the basis of a pluralistic conservation strategy, "bio" must be interpreted broadly to include both patterns and processes of nature as well as much of what is technically considered "abiotic." Maintaining dynamic patterns of biodiversity requires attention to underlying functions such as hydrological and climatological processes, nutrient cycles, disturbance regimes, dispersal of seeds and spores, and adaptation. Although these processes still operate at some level in impoverished ecosystems, their disruption is often at the root of impoverishment. The ESA and other environmental legislation, even if fully enforced, are inadequate to maintain the myriad patterns and processes of biodiversity—especially at higher levels of organization. Clearly, we need to think bigger.

WHICH SPECIES?

Certainly we are unlikely to change the species-centered bias of conservation overnight. Moreover, there are many practical reasons for retaining an interest in species. Instead of trying to assess the viability of an entire, complex ecosystem, we might concentrate on those species that play critical roles in the ecosystem. The problem then becomes one of selecting the most suitable species for detailed monitoring and viability analysis (Soulé, 1987). Five overlapping classes of species deserve special emphasis: indicators, keystones, umbrellas, flagships, and vulnerables. These species groups can be used to maintain the integrity of higher levels of biological organization. A species that falls into several of these groups would warrant extreme protection. More to the point of this book, such an analysis of species "importance" might form the basis of a priority system for listing and recovering endangered species.

Indicator Species

Of the five groups of species mentioned above, the indicators are probably the weakest for conservation purposes. Indicator species have been commonly used in environmental assessments. Taxa with known tolerances to specific pollutants can be used to evaluate air or water quality or to screen for subsequent chemical tests. This use of indicator species has limited applicability, however. Whenever ecological degradation is measured by only a few parameters, it can lead to fallacious conclusions. Ideally, environmental assessments should consider a range of ecological parameters at several levels of organization, integrated into some index of biotic integrity (Karr and others, 1986).

Indicator species also have been used to evaluate the effects of land management on wildlife. The National Forest Management Act of 1976 requires the U.S. Forest Service to describe the impact of alternative management plans on a set of "management indicator species" (MIS). Five categories of species are to be considered in choosing MIS: threatened and endangered species; species sensitive to intended management practices; game and commercial species; nongame species of special interest; and ecological indicator species that suggest the effects of management practices on a broad set of species (Salwasser and others, 1983; Wilcove, 1988). Species chosen as management indicators are usually vertebrates, but they can include invertebrates, plants, and even habitat types. The MIS concept is based on the assumption (often violated) that species are closely associated with structural features of their habitats.

As the politics of public land management now stand, the use of indicator species must be viewed with caution. Application of the indicator concept is subject to bias (Norse and others, 1986; Landres and others, 1988); species chosen as indicators may be edge-adapted or weedy species that prosper in heavily logged and fragmented forests. Species of old-growth and other vulnerable habitats have fared poorly under the indicator approach (Crawford and others, 1981; Graul and Miller, 1984; Noss and Harris, 1986). In many cases, management indicator species used by the Forest Service (such as the raccoon) are so adaptive to habitat change that their abundance tells little about habitat conditions. Imprudent implementation does not invalidate a concept, of course. Some species or guilds undoubtedly are useful indicators of management effects on their own or higher levels of organization. Ideally, indicators should be selected by biologists who are familiar with local and regional ecology but have no vested interest in resource uses such as timber production.

Keystone Species

A keystone species is a species that plays a pivotal role in an ecosystem and upon which the diversity of a large part of the community depends. Keystone species are often the dominant species at their trophic level, but their importance in a biotic community is greater than abundance alone might suggest. The keystone concept originated with Paine's (1966) demonstration that a predatory starfish enhanced diversity at lower trophic levels by preying preferentially on a dominant mussel herbivore that otherwise excluded inferior competitors. Subsequently, Gilbert (1980) showed that certain neotropical "keystone mutualist" plant species support many animal pollinators and seed dispersers and thus permit other plant species to persist. Keystone plant resources in southeastern Peru include palm nuts and figs, which sustain fruit-eating species during periods of food scarcity (Terborgh, 1986).

Animal species in a variety of communities are being identified as keystones. Beavers and pocket gophers maintain wetlands and meadows, respectively, which provide habitat for numerous plants and animals. Red-naped sapsuckers in Colorado create and maintain "wells" in shrubby willows upon which many herbivores feed (Ehrlich and Daily, 1988). Top predators often hold the key to the diversity and stability of ecosystems (Terborgh, 1988).

In longleaf pine ecosystems of the southeastern United States, the gopher tortoise digs burrows up to 30 feet long and 15 feet deep, in which some 362 species of invertebrates and vertebrates have been found. Some of these species are obligate commensals—that is, they are absolutely dependent upon the threatened tortoise. In the same system, a dominant and flammable groundcover plant, wiregrass, encourages frequent low-intensity ground fires. These fires maintain an open, parklike forest with one of the richest assemblages of herbaceous plants in the world (Clewell, 1989; Noss, 1988). Longleaf pine also plays a critical role in this process by converting lightning strikes into ground fires and by producing highly combustible duff (Platt and others, 1988). Animals such as the gopher tortoise, its many commensals, and the endangered red-cockaded woodpecker require the open habitat maintained by fire. At least fourteen species of cavity-nesting birds and mammals, in turn, are dependent upon the red-cockaded woodpecker which excavates cavities in living, old-growth pines. Thus an ecosystem may be characterized by several interdependent keystone species, all interacting with a keystone abiotic factor (in this case, fire).

Few biologists would dispute the rationale for devoting extra attention to keystone species; if such species are eliminated, a cascade of secondary extinctions may follow. An unanswered question is how "key" a species must be in its ecosystem to warrant extra attention. Identifying keystone species in every ecosystem worldwide (fundamentally a problem of the ancient but much neglected discipline of natural history) should be a high research priority.

Umbrellas and Flagships

Less scientific, but no less important, is the justification for protecting umbrella and flagship species. Umbrella species have large area requirements. Hence if enough contiguous areas are set aside to maintain their populations, many other species will be carried under the umbrella of protection. Large carnivores are obvious umbrellas (and are often keystones as well). The related category of flagship species is composed primarily of "charismatic megavertebrates" that serve as symbols for major conservation efforts. The Florida panther, for example, chosen by Florida schoolchildren as the state mammal, is a symbol of Florida's vanishing wilderness. If given enough roadless, wild habitat to maintain a viable population (which, at present, it is not), many other species would be protected. In Brazil, public awareness campaigns for two primates, the muriqui and the golden lion tamarin, have been very effective in increasing popular support for conservation of the Atlantic forest (Mittermeier, 1988).

Vulnerable Species

Finally, species that are highly vulnerable to the impact of human civilization—regardless of their ecological "importance"—must be protected if we are to maintain the full spectrum of biodiversity. It is only prudent to devote most conservation attention to those species most endangered by prevailing land use. A number of traits predispose species to decline in human-dominated landscapes: small population size, a low level of genetic variation, poor dispersal ability, large area requirements, low fecundity, dependence on patchy or unpredictable resources, a tendency to congregate in large groups, long-distance migration, and ground-nesting habits. (See Terborgh and Winter, 1980; Karr, 1982; Soulé, 1983; Pimm and others, 1988.) A species that possesses several of these characteristics may be particularly vulnerable. Vulnerability can also be estimated empirically by the extent or rapidity of decline since human settlement. Regional analyses of species vulnerability to land-use practices are needed to guide conservation agencies and land-use planners.

What About Rarity?

The reader may have noted that rarity—the traditional criterion of species conservation and the basis of the ESA—has not been emphasized here. Certainly rarity is a good predictor of vulnerability; small populations are prone to elimination from environmental randomness, catastrophe, demographic randomness, social dysfunction, genetic deterioration, and other problems (Shaffer, 1981; Soulé and Simberloff, 1986). But rarity is a poor indicator of ecological role or importance. An ecosystem is more endangered when keystone species decline substantially, but are not yet endangered (by standard criteria), than when the last few individuals are lost from species that were always rare.

Furthermore, not all rare species are vulnerable to further decline. Statistical studies of species abundances in diverse communities show that *most* species are relatively rare even in the absence of human disturbance. To evaluate risk of extinction, knowledge of rarity must be followed by consideration of life history and other factors that affect survival under specified conditions (Burke and Humphrey, 1987). Despite these problems, inventories of rare species (such as The Nature Conservancy's heritage programs) should be continued and intensified.

WILDERNESS AS A BASELINE

These arguments suggest that conservation should devote most (but not exclusive) attention to those species whose protection would encourage the persistence of higher levels of organization. A complementary approach is to focus directly on higher levels such as ecosystems. Indeed, there is an emerging consensus within the conservation community that our effort should be directed more toward saving habitat than managing individual species.

In the Pacific Northwest, where the spotted owl is considered an indicator of large expanses of old-growth forest, logging is restricted on some public lands inhabited by this rare species. Yet the Forest Service currently plans to protect only 14 percent of total spotted owl habitat on lands designated suitable for timber production (Wilcove, 1988). More than 40,000 acres of old-growth forest are logged each year (nearly 100,000 acres in 1990) from public lands in Oregon and Washington, and only about 10 percent of the original old growth remains (Wilderness Society, 1988). Politics has overridden biology in this debate, and protection for the owl is not assured. Even now that the spotted owl is listed, there is no assurance that the minimum habitat needed by owls is the minimum necessary to maintain the old-growth ecosystem with all its attendant species and processes. Listing the red-cockaded woodpecker, a species dependent on old-growth longleaf pines, has not prevented further devastation of longleaf pine forests on public lands in the Southeast. Nor has it permitted recovery of the woodpecker (Jackson, 1986).

How much more old growth could be saved if such forests, wherever they occur, were recognized as "endangered ecosystems" and protected accordingly? Several ecologists (among them Noss and Harris, 1986; Hunt, 1989) have discussed the need for an Endangered Ecosystems Act. Perhaps a better approach would be a Native Ecosystems Act that includes two components: endangered ecosystems and representative ecosystems. The first section would prohibit the "taking" (logging, road building, and so forth) of ecosystem types that have declined substantially (for example, by 80 percent or more) over their range and would set recovery goals for restoring these ecosystems. The second section would avoid such last-ditch efforts in the future by requiring that exemplary areas of various ecosystem types be set aside in large, intact units.

Large public land holdings offer virtually the only opportunity in this country to maintain representative ecosystems and the diverse features that define biodiversity at a landscape scale. Unfortunately, these lands are being increasingly fragmented by roads, clear-cuts, and other intrusions. Research natural areas (RNAs) and other small set-asides may suffice in the short term for plant species, restricted community types, and certain animal species. But small reserves are inadequate to protect wide-ranging animals. Moreover, they are unable to incorporate complete environmental gradients, disturbance regimes, and internal recolonization sources to stabilize patch dynamics (Pickett and Thompson, 1978).

Natural landscape patterns and processes are best maintained in large wilderness areas. These relatively unmodified landscapes should be recognized as scientific control areas with which multiple-use lands may be compared. How can we hope to evaluate the success of our management programs without a baseline of essentially unmanaged wilderness? Apart from their benchmark functions, wilderness areas maintain a reservoir of biodiversity as well as options for adaptation to an uncertain future. This is not to suggest that biodiversity can be relegated to wilderness areas and other set-asides. Soon there simply will not be enough land in set-asides to contain more than a sampling of biodiversity. Hence all lands—even those intensively used by humans—must be managed for biodiversity. But the value of set-asides increases with each additional acre subjected to human use.

Wilderness designation has been based primarily on scenic and recreational values, however, rather than scientific criteria. Moreover, areas that are highly productive in terms of timber usually are excluded from consideration. Thus the array of ecosystem types represented in our wilderness system is biased toward low-diversity lands such as alpine zones. Of some 261 ecosystem types defined by a combination of Bailey's ecoregions and Kuchler's potential natural vegetation, 104 (40 percent) are not represented in the National Wilderness Preservation System (Davis, 1988). Only fifty (19 percent) of these ecosystem types are represented in wilderness areas in units of at least 250,000 acres, half of them in Alaska. Clearly much more land, representing a larger array of ecosystem types, should be protected as wilderness. Existing wilderness boundaries need to be enlarged, and unnatural disturbances such as livestock should be removed. Where legally definable wilderness is lacking, "wilderness recovery areas" can be created by closing roads, actively restoring habitats and natural disturbance regimes, and reintroducing extirpated species (Noss, 1987, 1988).

When Aldo Leopold first advocated wilderness designation in 1920, he spoke purely in terms of recreation. But by the mid-1930s, Leopold's knowledge and vision had matured and he emphasized the scientific and ecological value of wilderness. In 1941, Leopold wrote of wilderness as vital to "the science of land health" because it offered "a base-datum of normality, a picture of how healthy land maintains itself as an organism" (Leopold, 1941, p. 3).

TOMORROW?

Conservation was simpler when all we had to worry about were such straightforward problems as egret plumes on ladies' hats. Then came the Dust Bowl, silted streams, and belated recognition of the effects of the plow, the axe, and the cow on once healthy land. By the 1960s and 1970s, we had DDT, acid rain, toxic wastes, habitat fragmentation, and edge effects to trouble us. Lists of endangered species swelled. Today, in addition to these threats, we face the specter of ozone holes, deadly smog, and global warming brought on by our combustion of fossil fuels and massive deforestation. Underlying these ills is human overpopulation and profligate consumption. Worse yet, environmental threats interact in a multiplicative, synergistic fashion with unpredictable effects that have no easy remedy. Even the most remote wilderness, our "base-datum of normality," is not safe from these global perils. Yet a shifting baseline is better than no baseline at all.

Given the immensity of the biodiversity crisis and the uncertainty of the cause-effect relationships underlying it, what practical steps can we take to lessen the impacts? One key principle is prudence. In other words, we should err on the side of preservation when science is unable to provide clear answers. Reversing human population growth and resource consumption are long-term necessities. Scientifically credible environmental education must become a prominent feature of school curricula. Because the threats to biodiversity are so immediate—the battle may be won or lost in the 1990s—our imperative is action. Conservation biology is a contentious science, true, but there is an emerging consensus on certain elements of an active macrostrategy.

IDENTIFY HOT SPOTS

Although work toward a complete biological inventory should continue, what we need now are quick surveys of existing data to delineate hot spots of species richness and endemism. (See Scott and colleagues in Part IV of this volume.) Hot spots should be stratified by biogeographic region in order to represent each ecosystem type and each flora and fauna fairly. Tundra is no less precious than rainforest simply because it contains fewer species. Global biodiversity depends on adequate representation of all ecosystem types.

ANALYZE THREATS AND VULNERABILITY

Although we probably know the major threats to biodiversity, we know very little about specific impacts. Research to this end is needed. Also needed are vulnerability analyses at several levels of biological organization. We need to identify the species, habitats, and regions most sensitive to human pressure, as well as the regions subjected to the highest levels of stress. These data should be overlaid with data on species richness and endemism to develop a set of priorities for conservation action. Regions of high species richness, high endemism, high sensitivity, and high levels of stress warrant the quickest protective action.

PROTECT, RESTORE, MANAGE, AND REGULATE

Hot spots should be acquired and protected as reserves; the most vulnerable areas should be targeted for immediate purchase. Multiple reserves are needed to protect genetically distinct populations of species and to avoid contagious catastrophes. Only the largest reserves, however, can be expected to encompass the shifting mosaic of disturbance-recovery processes and the home ranges of large carnivores and migratory ungulates. No reserve is immune to climate change and atmospheric abuse. Regional networks of interconnected reserves, integrated into the surrounding multiple-use landscape by a gradation of buffer zones, allow for flexibility in a changing world. (See Noss, 1983, 1987; Harris, 1984; Noss and Harris, 1986; Salwasser and others, 1987.) Maintaining a full range of physical habitats within protected areas and corridors will provide opportunities for species to track shifting climates over time (Hunter and others, 1988).

Broad, continental-scale habitat corridors will be particularly useful as an avenue for species dispersal during climate change, although the predicted rates of change will make migration difficult for the less mobile species. Translocations of species, as well as intensive habitat restoration, preparation, and management activities, will be increasingly necessary (Franklin and others, 1988). Deforestation must be stopped, and extensive areas must be reforested to sequester carbon from the atmosphere. The seminatural matrix of nonurban, nonagricultural land, outside of strict reserves, will become more important as a reservoir of biodiversity and ecological function; hence land-use planning must be expanded to regional, national, and international scales (Brown, 1988). Land-use and human settlement patterns must be regulated, much more so than today, if we are to optimize the contribution of each land parcel to regional and global conservation strategies.

BE EXPEDIENT

We need to apply vigorously what tools we have, even when they are not ideal for the job. At present, we have the Endangered Species Act and a wealth of other imperfect environmental legislation. These laws must be enforced without compromise. Circumvention of the ESA, as in the case of the Mt. Graham red squirrel, must be fought to the end. As we move into the next decade of conservation, we must develop stronger tools (such as a Native Ecosystems Act) and greater public awareness of the problems.

BE VISIONARY

The biodiversity challenge extends far beyond law, science, and conventional considerations of public policy. The challenge is to develop what Aldo Leopold called a land ethic. When entire faunas and floras are disintegrating and fundamental ecological processes are breaking down, the only tenable response is a drastic change in the way we interact with the earth. Biodiversity is no frill—it is life and all that sustains life. It is worthy of respect. Maintenance of biodiversity must become our primary mission as a society, the principle that guides all resource uses. The haunting question is: Will public awareness of the crisis catch up in time? We cannot afford to wait for an answer.

Note

I thank Blair Csuti, Ed Grumbine, Ann Hairston, Sandra Henderson, Kathryn Kohm, Thom Whittier, and David Wilcove for their helpful comments on earlier drafts of this essay.

References

Allen, T.F.H., and T. B. Starr. 1982. *Hierarchy: Perspectives for Ecological Complexity.* Chicago: University of Chicago Press.

Brown, J. H. 1988. "Alternative Conservation Priorities and Practices." Paper presented at 73rd annual meeting, Ecological Society of America, Davis, California.

Burke, R. L., and S. R. Humphrey. 1987. "Rarity as a Criterion for Endangerment in Florida's Fauna." *Oryx* 21:97–102.

Clewell, A. F. 1989. "Natural History of Wiregrass (*Aristida stricta* Michx., Gramineae)." *Natural Areas Journal* 9(4):223.

Crawford, H. S., R. G. Hooper, and R. W. Titterington. 1981. "Songbird Population Response to Silvicultural Practices in Central Appalachian Hardwoods." *Journal of Wildlife Management* 45:680–692.

Davis, G. D. 1988. "Preservation of Natural Diversity: The Role of Ecosystem Representation Within Wilderness." Paper presented at National Wilderness Colloquium, Tampa, Florida, January 1988.

Ehrlich, P. R., and G. C. Daily. 1988. "Red-Naped Sapsuckers Feeding at Willows: Possible Keystone Herbivores." *American Birds* 42:357–365.

Franklin, J. F., K. Cromack, W. Denison, A. Mckee, C. Maser, J. Sedell, F. Swanson, and G. Juday. 1981. *Ecological Characteristics of Old-Growth Douglas-Fir Forests.* USDA Forest Service General Technical Report PNW-118. Portland: Pacific Northwest Forest and Range Experiment Station.

Franklin, J., V. H. Dale, D. Perry, F. J. Swanson, M. Harmon, T. A. Spies, A. McKee, and D. Larson. 1988. "Effects of Global Climatic Change on Forests in Northwestern North America." Paper presented at World Wildlife Fund Conference on Consequences of the Greenhouse Effect for Biological Diversity, Washington, D.C.

Gilbert, L. E. 1980. "Food Web Organization and the Conservation of Neotropical Diversity." Pp. 11–33 in M. E. Soulé and B. A. Wilcox (eds.), *Conservation Biology: An Evolutionary-Ecological Perspective.* Sunderland, Mass.: Sinauer Associates.

Graul, W. D., and G. C. Miller. 1984. "Strengthening Ecosystem Management Approaches." *Wildlife Society Bulletin* 12:282–289.

Harris, L. D. 1984. *The Fragmented Forest: Island Biogeography Theory and the Preservation of Biotic Diversity.* Chicago: University of Chicago Press.

Hunt, C. E. 1989. "Creating an Endangered Ecosystems Act." *Endangered Species Update* 6(3–4):1–5.

Hunter, M. L., G. L. Jacobson, and T. Webb. 1988. "Paleoecology and the Coarse-Filter Approach to Maintaining Biological Diversity." *Conservation Biology* 2:375–385.

Hutto, R. L., S. Reel, and P. B. Landres. 1987. "A Critical Evaluation of the Species Approach to Biological Conservation." *Endangered Species Update* 4(12):1–4.

Jackson, J. A. 1986. "Biopolitics, Management of Federal Lands, and the Conservation of the Red-Cockaded Woodpecker." *American Birds* 26(40):1162–1168.

Karr, J. R. 1982. "Population Viability and Extinction in the Avifauna of a Tropical Land Bridge Island." *Ecology* 63:1975–1978.

Karr, J. R., K. D. Fausch, P. L. Angermeier, P. R. Yant, and I. J. Schlosser. 1986. *Assessing Biological Integrity in Running Waters: A Method and Its Rationale.* Special Publication No. 5. Champaign: Illinois Natural History Survey.

Kellert, S. R. 1980a. "Contemporary Values of Wildlife in American Society." Pp. 31–61 in W. W. Shaw and E. H. Zube (eds.), *Wildlife Values.* Fort Collins, Colo.: U.S. Forest Service Center for Assessment of Noncommodity Natural Resource Values.

Kellert, S. R. 1980b. "Americans' Attitudes and Knowledge of Animals." *Transactions of the North American Wildlife and Natural Resources Conference* 45:111–124.

Koestler, A. 1967. *The Ghost in the Machine.* New York: Macmillan.

Kushlan, J. A. 1979. "Design and Management of Continental Wildlife Reserves: Lessons from the Everglades." *Biological Conservation* 15:281–290.

Landres, P. B., J. Verner, and J. W. Thomas. 1988. "Ecological Uses of Vertebrate Indicator Species: A Critique." *Conservation Biology* 2:316–328.

Leopold, A. 1941. "Wilderness as a Land Laboratory." *Living Wilderness* 6(July):3.

Leopold, A. 1949. *A Sand County Almanac and Sketches Here and There.* Oxford: Oxford University Press.

Meine, C. 1988. *Aldo Leopold: His Life and Work.* Madison: University of Wisconsin Press.

Mittermeier, R. A. 1988. "Primate Diversity and the Tropical Forest: Case Studies from Brazil and Madagascar and the Importance of the Megadiversity Countries." Pp. 145–154 in E. O. Wilson (ed.), *Biodiversity.* Washington, D.C.: National Academy Press.

Newmark, W. D. 1985. "Legal and Biotic Boundaries of Western North American National Parks: A Problem of Congruence." *Biological Conservation* 33:197–208.

Norse, E. A., K. L. Rosenbaum, D. S. Wilcove, B. A. Wilcox, W. H. Romme, D. W. Johnston, and M. L. Stout. 1986. *Conserving Biological Diversity in Our National Forests.* Washington, D.C.: Wilderness Society.

Noss, R. F. 1983. "A Regional Landscape Approach to Maintain Diversity." *BioScience* 33:700–706.

Noss, R. F. 1987. "Protecting Natural Areas in Fragmented Landscapes." *Natural Areas Journal* 7:2–13.

Noss, R. F. 1988. "The Longleaf Pine Landscape of the Southeast: Almost Gone and Almost Forgotten." *Endangered Species Update* 5(5):1–8.

Noss, R. F., and L. D. Harris. 1986. "Nodes, Networks, and MUMs: Preserving Diversity at All Scales." *Environmental Management* 10:299–309.

Office of Technology Assessment (OTA). 1987. *Technologies to Maintain Biological Diversity,* OTA-F-330. Washington, D.C.: U.S. Gov't Printing Office.

O'Neill, R. V., D. L. DeAngelis, J. B. Waide, and T.F.H. Allen. 1986. *A Hierarchical Concept of Ecosystems.* Princeton, N.J.: Princeton University Press.

Paine, R. T. 1966. "Food Web Complexity and Species Diversity." *American Naturalist* 100:65–75.

Pickett, S.T.A., and J. N. Thompson. 1978. "Patch Dynamics and the Design of Nature Reserves." *Biological Conservation* 13:27–37.

Pimm, S. L., H. L. Jones, and J. Diamond. 1988. "On the Risk of Extinction." *American Naturalist* 132:757–785.

Platt, W. J., G. W. Evans, and S. L. Rathbun. 1988. "The Population Dynamics of a Long-Lived Conifer (*Pinus palustris*)." *American Naturalist* 131:491–525.

Salwasser, H., C. Schonewald-Cox, and R. Baker. 1987. "The Role of Interagency Cooperation in Managing for Viable Populations." Pp. 159–173 in M. E. Soulé (ed.), *Viable Populations for Conservation.* Cambridge: Cambridge University Press.

Salwasser, H., C. K. Hamilton, W. B. Krohn, J. F. Lipscomb, and C. H. Thomas. 1983. "Monitoring Wildlife and Fish: Mandates and Their Implications." *Transactions of the North American Wildlife and Natural Resources Conference* 48:297–307.

Shaffer, M. L. 1981. "Minimum Population Sizes for Species Conservation." *BioScience* 31:131–134.

Sidle, J. G., and D. B. Bowman. 1988. "Habitat Protection under the Endangered Species Act." *Conservation Biology* 2:116–118.

Soulé, M. E. 1983. "What Do We Really Know About Extinction?" Pp. 111–124 in C. M. Schonewald-Cox and others (eds.), *Genetics and Conservation: A Reference for Managing Wild Animal and Plant Populations.* Menlo Park, Calif.: Benjamin/Cummings.

Soulé, M. E. 1987. "Introduction." Pp. 1–10 in M. E. Soulé (ed.), *Viable Populations for Conservation.* Cambridge: Cambridge University Press.

Soulé, M. E., and D. Simberloff. 1986. "What Do Genetics and Ecology Tell Us About the Design of Nature Reserves?" *Biological Conservation* 35:19–40.

Terborgh, J. 1986. "Keystone Plant Resources in the Tropical Forest." Pp. 330–344 in M. E. Soulé (ed.), *Conservation Biology: The Science of Scarcity and Diversity.* Sunderland, Mass.: Sinauer Associates.

Terborgh, J. 1988. "The Big Things That Run the World—A Sequel to E. O. Wilson." *Conservation Biology* 2:402–403.

Terborgh, J., and B. Winter. 1980. "Some Causes of Extinction." Pp. 119–133 in M. E. Soulé and B. A. Wilcox (eds.), *Conservation Biology: An Evolutionary-Ecological Perspective.* Sunderland, Mass.: Sinauer Associates.

Wilcove, D. S. 1988. *National Forests: Policies for the Future.* Vol. 2: *Protecting Biological Diversity.* Washington D.C.: Wilderness Society.

Wilcove, D. S., C. H. McLellan, and A. P. Dobson. 1986. "Habitat Fragmentation in the Temperate Zone." Pp. 119–133 in M. E. Soulé (ed.), *Conservation Biology: The Science of Scarcity and Diversity.* Sunderland, Mass.: Sinauer Associates.

Wilderness Society. 1988. "Litigation Update." *Wilderness* 52(183):6–9.

IN SEARCH OF AN ECOSYSTEM APPROACH TO ENDANGERED SPECIES CONSERVATION

by

HAL SALWASSER

FOLLOWING PASSAGE of the Endangered Species Act in 1973, this country embarked on a journey to protect what seemed to be a manageable number of endangered species. We designated critical habitats and limited the use of harmful chemicals such as DDT. In 1990, just a decade and a half later, we see that the magnitude of the endangered species problem is far greater than originally imagined. At last count, the federal endangered species list had topped 1,000 species with over 3,000 species waiting as candidates for listing. Moreover, states have listed even more species as endangered. In short, biological evidence shows we are only scratching the surface of a larger extinction crisis.

If the lessons of more than one hundred years of wildlife conservation are correct, including those from the field of conservation biology, population viability is likely to be in jeopardy for more species than will ever be formally listed by agencies of government. This finding, coupled with the growing number of listed species that need immediate recovery measures, compels something beyond crisis management of each species as it appears on the brink of extinction. Many scientists and resource managers speak of that "something beyond" as ecosystem management. This essay explores ecosystem management for conservation of endangered species and suggests some actions needed to implement such a strategy.[1]

First we must dispense with the simplistic notion that endangered species can be preserved if we just have more protected natural areas. The world does not have, nor is it ever likely to have, enough parks, refuges, and nature preserves to protect more than a small fraction of the currently recognized endangered species, if only such areas are employed in that cause. These highly protected lands currently total about 3 percent of the world's surface area (World Resources Institute, 1989). Even doubling the number or size of nature preserves in the world—a politically optimistic prospect to say the least—would still leave us far short of what would be needed to protect all current or future endangered species.

Ecosystems at geographic scales where it is meaningful to consider sustaining the richness and variety of life are typically on the order of millions of hectares. Such large areas inevitably reflect past and current human activities and will be expected to provide a variety of natural resources for people. This sets up the ultimate challenge in conservation: The future for endangered species, as for all of life's variety, will be determined by how quickly and how well we learn to integrate goals for a rich biotic future with our growing need for resources for subsistence, commerce, recreation, and spiritual renewal. So-called multiple-use lands, where a balance is struck between protection of natural systems and production of resources, will be increasingly important. It would help if there were fewer humans, and if people were not so prodigal in their consumption of raw materials. But there aren't, and they aren't, and we must deal with these realities.

We also need to augment the current "one at a time" strategy of recovery and rehabilitation of endangered species. We need a more encompassing strategy for conserving biological diversity—a strategy whose long-term success depends on conservation actions that protect genetic resources, sustain population viability of all species, perpetuate natural biological communities, and maintain a full range of ecological processes while meeting human needs. The customary approach of deciding what must be preserved and segregating it from the rest can have only limited results. It reflects a siege mentality that derives from the doom and gloom rhetoric on extinctions and biological diversity. The future need not be that hopeless. But it will be if conservationists do not learn to be more effective in achieving their goals in more and larger areas of land and water, and in ways compatible with other human uses for those lands. Here I suggest one approach to conserving endangered species through the broader goal of conserving biodiversity in large regional ecosystems. This framework draws from practical experience, empirical knowledge, and ecological theories I have found useful in managing forests and their wildlife.

BEYOND THE SIEGE MENTALITY

We have traditionally approached endangered species conservation one species at a time. This strategy results in much attention to a few species—the "charismatic species" such as bald eagles, grizzly bears, spotted owls, and gray wolves—and very little attention to most others. It also leads to a siege mentality in which the world is perceived to be closing in on rare species and their habitats. While this may be true in many cases, such a view of conservation causes people to approach species and habitat protection by trying to segregate them from human activities.

The upshot of a siege mentality on endangered species is that society begins to perceive parks, wilderness areas, and nature reserves as the only way to care for endangered species habitats. Areas containing endangered species are placed in land uses that emphasize their protection while the managers of the vast majority of other lands are relieved of all responsibility for endangered species and biological diversity. Such a trend, carried to its logical extremes, would eventually result in two options: Either endangered species will dominate land-use goals on a larger portion of the nation's land, including private holdings, or they will become increasingly perceived as a threat to the majority of wildland uses and values and therefore isolated from them.

Neither extreme is necessary in all cases. We need conservation approaches that do not leave societies with either/or choices. Human population growth increases competition between different interests for land uses. Such competition, when viewed as a zero-sum game (somebody wins, somebody loses), jeopardizes prospects for integration of endangered species and biodiversity goals into ecosystem management. If we fail to develop approaches that blend endangered species conservation into an overall strategy for conserving biodiversity as an integral part of meeting long-term human needs, growing human populations will be increasingly forced to choose the second option mentioned above. We need alternatives to the siege mentality that segregates endangered species from a human-dominated world.

EXTENDING A LAND ETHIC

Highly protected areas, such as parks, natural areas, and wilderness areas, will always be vital to a robust conservation effort. But we need a paradigm that builds on a more holistic view of resource management. One step in building this paradigm is to instill a land ethic as the philosophical foundation for conservation programs and policies—at least for those that affect lands which are still wild and managed for a variety of uses and values.

The biocultural land ethic proposed by Aldo Leopold (1949) is an excellent step toward building such a foundation. The American land ethic is not entirely consistent, but it tends toward what Leopold described as a state of harmony between people and land. Such a land ethic treats people as an integral part of the land system, both recipients and stewards of its richness. Leopold considered land to be a community of soils, waters, biota, and the ecological processes that hold them together. In this ecosystem view of nature, stewardship is the concept Leopold offered to characterize a sound human/land relationship.

Land stewardship means keeping biotic systems in good working order. Hence the first precaution for managers is to save "all the cogs and wheels." A principle of biological stewardship is to assure that human actions do not cause things to disappear before their time and that native species and biological communities continue to play out their evolutionary drama in as great a number and in as many places as possible. It seems that this is the essential goal underlying the biological mandates of the National Environmental Policy Act of 1969, the Endangered Species Act of 1973, as amended, and the National Forest Management Act of 1976.

Experience has shown that successful strategies to protect endangered species blend preservation goals with other human-centered considerations for what should happen to the soil, air, water, and biota in a particular place. There is a constant balancing between the mandate and means to protect biological resources and the desire to proceed with various developments to meet other social and personal goals. This balancing is central to a land ethic that harmonizes goals for natural resources with goals for healthy natural systems. It is also a key to implementing an ecosystem approach to conservation of endangered species.

FOCUSING ON SPECIFIC ECOSYSTEMS AND ELEMENTS OF DIVERSITY

Once we begin to think of the challenge of protecting endangered species as a matter of integrating diverse values and goals, we are faced with a number of practical problems. Not the least of these is finding ecologically and politically meaningful scales to focus our efforts. On the one hand, nature is full of gradients. On the other, humans like to classify things. Hence we develop rules for deciding what constitute species, biological communities, and ecosystems, even though we know that nature is rarely that distinct. It is practical to think of ecosystems as places defined by dominant physical or ecological factors, such as forests, lakes, watersheds, mountain ranges, and deserts. Any such human-defined ecosystem is a subset of a larger ecosystem, of course, up to and including the biosphere.

For conservation purposes, ecosystems must have both biological and practical applications. An ecosystem might be a relatively small area, such as the range of a rare plant, or as large as the geographic range of the peregrine falcon in the Western Hemisphere. The main point is that an ecosystem provides a meaningful context for addressing genetic variation, viability of species populations, continued existence of biological communities, and richness of ecological processes.

Ecosystem management is the care and tending of these human-defined areas to meet a variety of goals. It draws upon the natural restorative powers of the land, protects the diversity of natural communities, and provides resilience in the face of external forces. It is a hard concept to define and even harder to prescribe. However, it differs from traditional approaches to land and resource management such as forestry, fisheries, and wildlife habitat management in that it incorporates a wider breadth of ecological and social concerns as well as a longer time scale for productivity and renewability.

Any approach to conserving biodiversity through ecosystem management must focus on selected elements of the biota to guide planning, analysis, decision making, and monitoring. It is impossible to plan, analyze, and take action simultaneously on the thousands of species and communities likely to occur in an ecosystem of several hundred square kilometers. Nor is it rational just to let nature take its course. There is no reason to assume that nature will automatically recover endangered species or produce either the biodiversity or resources desired from an area (Cairns, 1989), especially if the area is relatively small.

The focal elements for ecosystem management (Figure 3) should vary with geographic scale and typically include a mix of individual species (perhaps ecological indicators or featured species), groups of species (such as management guilds), distinct biological communities (such as wetlands, snag-dependent wildlife, or old-growth forests), and landscape attributes (such as patterns and connectedness). At the geographic scale of a region, state, country, or continent, for example, appropriate focal elements for a biodiversity conservation strategy might include:

- Wide-ranging endangered species for which recovery is the goal (peregrine falcon, gray wolf)
- Wide-ranging species for which extinction is not a pressing issue but for which viability of well-distributed populations is a concern (spotted owl)
- Species of major commercial or recreational value for which genetic and demographic conditions needed for resilience and productivity are goals (Douglas fir, chinook salmon, elk)

FIGURE 3

FOCAL ELEMENTS FOR ECOSYSTEM MANAGEMENT

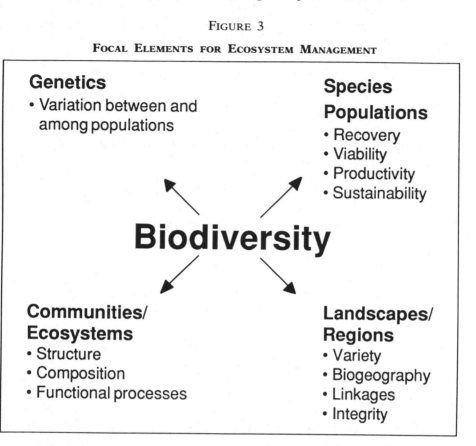

- Groups of species sensitive to environmental trends (cavity-dependent wildlife)
- Special habitats that are rare or declining in the region (old-growth forests, riparian areas, bottomland hardwoods)
- Key factors deemed essential to ecosystem function (landscape linkages, air and water quality)

At smaller scales such as national parks, national forests, watersheds, stands, or sites, the focal elements should reflect local or regional concerns.

The purpose of using focal elements is to bring immediate attention to those species, communities, and ecosystem attributes most in need of management action. It is unlikely that a set of focal elements small enough to be feasible for planning (for example, two to three dozen species, communities, and attributes at a particular geographic scale) can ever be used to infer that all species and processes in ecosystems are addressed. Thus focal elements should not be confused with the elusive concept of ecological indicators as discussed by Landres and colleagues (1988). What is important to an ecosystem approach for endangered species conservation is that species already considered by state and federal agencies to be endangered should be among the elements of diversity used for planning and carrying out ecosystem management.

ADOPTING PLANS THAT INTEGRATE GEOGRAPHIC SCALES

Integrating endangered species into ecosystem management will require strong, clear, consistent, and feasible policies for agencies and landowners at several administrative and geographic scales (Salwasser and others, 1987; Salwasser, 1988b). To encourage positive and innovative actions for endangered species conservation, coordination of policies and plans should not be approached as a punitive endeavor. The intent of plans and policies for biological diversity should be to empower positive action, not to compel bitter compliance. The approach taken by federal and state agencies in the Interagency Grizzly Bear Committee is a good example of the kind of coordination needed (Servheen, 1984). All relevant agencies willingly cooperate in carrying out combined programs in research, management, education, and monitoring that in turn help each agency fulfill its conservation mandate.

On-the-ground accomplishments require two basic elements: *specificity of direction* (what exactly is to happen, by whom, when, where, and with what expected result?) and *accountability for results* (how will achievement be measured and rewarded or how will failure be dealt with?). Specificity and accountability for endangered species conservation as well as overall biodiversity must be clear for at least four geographic scales (Figure 4). The larger scales provide general guidance to the smaller scales, but ultimate success depends on simultaneous accomplishment at all scales.

The largest scale is a region or the geographic range of focal

FIGURE 4

INTEGRATION OF GEOGRAPHIC SCALES

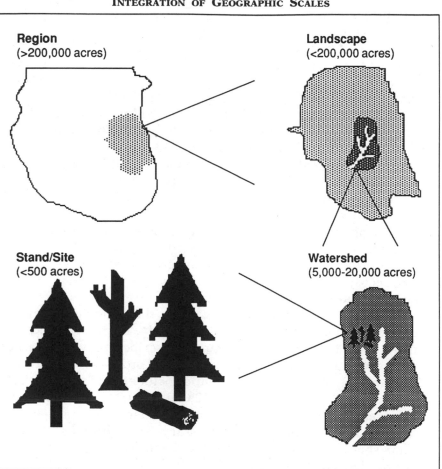

species or widespread biological communities of particular interest. For large or wide-ranging species such as grizzly bears, wolves, spotted owls, monarch butterflies, or peregrine falcons, this area could easily be a major portion of a continent or hemisphere. Plans at this scale should allocate lands to management strategies that favor different aspects of biodiversity. Designation of ecosystems where grizzly bears will be conserved in the continental United States, migration refuges for trumpeter swans, colony sites for red-cockaded woodpeckers, the habitat network for spotted owls, and the legal mandates for conservation systems such as national parks and national forests are examples of ecosystem approaches to biodiversity at large geographic scales.

The second geographic scale serves administrative as well as biological purposes. This is the scale at which administrative unit plans are developed and carried out or in which populations of wide-ranging species occur. Administrative units such as national forests, national parks, and wildlife refuges, or biological units such as demes, ungulate herds, flocks of migratory birds, and wolf packs, require attention at this scale. In addition, guidelines for focal elements of diversity should be integrated with other goals for large land areas. Allocation of sites or watersheds to management prescriptions that benefit various elements of diversity need to be specified, and overall plans for monitoring and evaluation should be proposed. Environmental assessments at this scale should describe the cumulative effects of policies and major resource management strategies on desired ecosystem attributes, communities, or species populations—that is, on the focal elements of diversity.

The third geographic scale draws on the ecological characteristics needed to meet goals for diversity elements at scales smaller than the administrative units described above. For large species with seasonal home ranges on the order of hundreds or thousands of acres, this scale would include units of land that are mappable as watersheds of 12,000 to 50,000 acres. Within such areas, habitat conditions can be managed and protection from humans provided where necessary. Environmental assessments should describe the cumulative effects of planned activities on desired attributes of biological communities, selected habitats, or subsets of species populations such as breeding pairs or social groups.

The fourth geographic scale includes mappable stands or sites that constitute individual elements of a landscape (that is, patches or microsites). For large or wide-ranging species, these sites may meet only part of their needs. Nevertheless, the structure, size, shape, location, and interconnectedness of such sites are crucial to the overall value of a landscape for many species. Although management prescriptions are commonly prepared for individual stands or sites, environmental assessments rarely would focus on such a small area.

These four geographic scales are not meant to be exhaustive. The scales used in ecosystem management cannot be standardized for all species or all environments. Plants, invertebrates, and small vertebrates may require different approaches. A plant whose entire range is within one meadow, for example, may not require attention at the first two scales. It may only need planning by a single landowner or administrative unit at the watershed scale and a specific management prescription at the stand scale. But a migratory bird that summers in Canada and winters in Mexico would require attention at all four scales, including an international component.

COORDINATING INTERAGENCY ACTIONS

Rarely, if ever, will a single administrative agency or landowner command an area so large that all aspects of biodiversity in a regional ecosystem are under one jurisdiction. Thus agencies concerned with wide-ranging species or plant communities are placed in a position of interdependence. Well-known examples include relationships between state wildlife and fisheries management agencies and the managers of wildlife and fish habitats, managers of wetlands along continental waterfowl flyways, managers of the genetic resources of anadromous fish and commercial tree species, and managers of the different seasonal habitats for migratory ungulates and birds.

The purpose of coordinating policies, goals, and actions between agencies and owners of shared resources is to enlarge the effective size and quality of the ecosystem for these resources. If national parks are isolated islands of habitat embedded in a matrix of inhospitable areas as postulated by Newmark (1987) and others, island biogeographic theory predicts that extinction rates in the parks will be inversely proportional to their size and time since isolation. In fact, many species would already have gone extinct in smaller or older parks. That this has not happened as predicted by theory (Quinn and others, in press) shows that parks and their surrounding public wildlands, when managed to sustain native species, communities, and ecological processes, create effectively larger ecosystems than any single unit can provide.

The continued high diversity of native species in national

parks and other public lands is not a result of specific goals for coordination of policies on biological diversity. It is an artifact of relatively infrequent or low intensities of human disturbance and the general conservation approaches historically employed on public wildlands—namely multiple-use, sustained-yield resource management. In the future, we may not be able to rely upon such general strategies because pressure for all land uses is intensifying. It may become necessary to actively coordinate ecosystem management strategies. The goal of such coordinated strategies should be to ensure that sufficient amounts of different habitats are maintained in patterns and connected so that species function as if a complex of many landownerships and agency jurisdictions were one contiguous ecosystem. As shown by Figure 5, the prospects for this, at least in the western United States, are still quite good.

There are several ways to achieve the kind of interagency coordination needed for regional ecosystem management. While there may be no preferred approach for all ecosystems, it is likely that some combination of the following models will be useful:

- Verbal agreements between neighboring resource managers. These occur frequently and are the easiest to make and the easiest to break. They work well for most issues.
- Formal memorandums of understanding or interagency agreements that provide for sharing of resources and decision making. Examples include memorandums of understanding between the Forest Service and forty-three state wildlife agencies, the Interagency Grizzly Bear Committee charter, and the Interagency Agreement on Spotted Owl Management. Although these agreements are more difficult to forge, they are also more durable and enforceable.
- Biosphere reserves (still more a concept than a tested practice).
- Congressionally or administratively chartered interagency mandates to achieve specified objectives, perhaps through specified mechanisms. An example is the interagency approach on wild horses and burros chartered by Congress. These arrangements are the most difficult to develop as well as the most binding and enforceable.

FIGURE 5

POTENTIAL BIOREGIONS

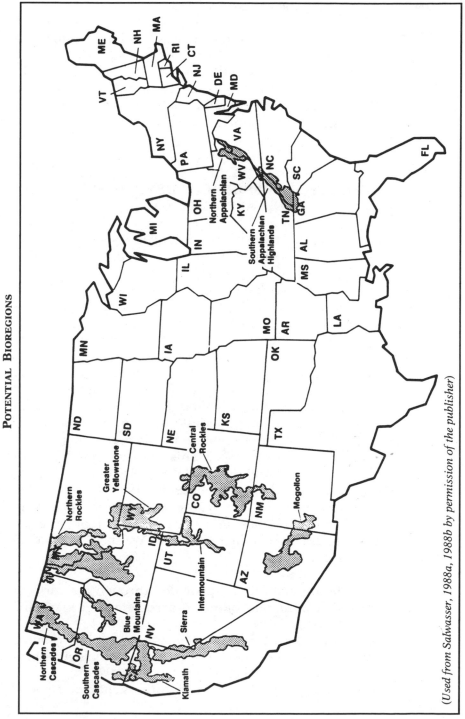

(*Used from Salwasser, 1988a, 1988b by permission of the publisher*)

- Congressional action to join disparate land units into a single administrative entity. A hypothetical example might be placing all federal lands in the Sierra Nevada range in California under a single agency's jurisdiction. It is now administered by units of the National Park Service, Forest Service, Bureau of Land Management, and numerous state and private entities.

INTEGRATING RESEARCH, DEVELOPMENT, AND APPLICATION

The knowledge needed to integrate endangered species into ecosystem management is different from that needed to determine how much land is required to segregate endangered species from humans. Scientists have concentrated more on the latter than the former. For example, the concepts of minimum critical ecosystem size (Lovejoy and Oren, 1981), minimum dynamic area (Pickett and Thompson, 1978), boundary effects (Schonewald-Cox and Bayless, in press), and treating parks as island faunas (Newmark, 1987) focus on questions of how big reserve areas should be to sustain their flora and fauna and what characteristics their boundaries should have. These are useful studies, but they beg the fundamental question: How should we manage large areas of land to sustain biodiversity and produce the resources desired by people? This question has been approached by Thomas (1979), Harris (1984), Salwasser, Mealey, and Johnson (1984), and Salwasser (1988a, 1988b).

Unless society intends to relegate endangered species to parks, refuges, and wilderness areas, we will need more integrative R&D than has been conducted to date. We should be asking questions such as:

- Can population viability be provided through networks of partially discontinuous habitats across large geographic areas?
- Can old-growth forests be managed as a renewable resource for perpetual yields of their biotic richness, ecological processes, and special wood products?
- Can management for ecological conditions characteristic of

old-growth forests (or, in the general sense, mature ecosystems) be integrated in landscapes that also have areas under intensive silviculture for wood fiber production (or other resources) and some areas under long-term experimental research?

One of the best models for integrating research, technology development, and actual conservation applications is the "adaptive environmental assessment and management" approach developed by Holling (1978), Walters (1986), and their colleagues. This approach recognizes that many significant problems in conservation are not amenable to traditional research: They are simply too big, too complex, too variable, and too long-term. The adaptive management approach offers a model for developing management plans as long-term experiments using the best data and analytic methods available. Monitoring and research are then used to learn from the "experiment" of resource management. Instead of the traditional separation of science from management, the two are long-term partners in a mutually beneficial endeavor.

CAPITALIZING ON CONSTITUENT SUPPORT

Finally, ecosystem approaches to conserving endangered species as integral parts of biological diversity will require greater public awareness, commitment, and political support. These all depend on information—who has it and what they do with it. Scientific information in technical journals is essential but not sufficient for success. Popular information, widely available and packaged to command attention and action, also is needed. Laws, regulations, and litigation have helped gain attention, but they have not created universal success for endangered species conservation. The same can be said for advances in scientific knowledge.

The common thread running through endangered species success stories is effective communication of issues, needs, knowledge, and the benefits of conservation. Such communication captures widespread support for a program. The effectiveness of the Peregrine Fund is an excellent illustration of this point;

similar efforts on behalf of grizzly bears in the Yellowstone area demonstrate the importance of communication and education in conservation programs.

Integrating endangered species into ecosystem management for biodiversity and sustainable resource development may mean that some of the glamour of single-species approaches will be lost and that resources will become correspondingly more difficult to secure. We may need to borrow concepts of effective marketing from the business world to gain that essential support.

FROM VISION TO REALITY

To accomplish the formidable conservation tasks before us, sustaining biological diversity must become a basic goal of land and resource management in as many areas of the world as is socially and economically possible. Conservation of endangered species can become part of an overall policy for ecosystem management that sustains biodiversity, perhaps the fundamental part. But that means the goals for endangered species must be blended with other goals for land use. To make that happen, there must be a significant shift in our attitudes toward endangered species problems. Defensive and regulatory strategies that derive from a siege mentality should become the strategies of last recourse in favor of positive approaches to integrating endangered species goals into ecosystem management. Thus conservationists must develop skills in integrated, multiuse resource management and negotiation.

A competitive edge for endangered species managers requires not only knowledge and technology but also the political and economic strength that comes from a large and influential constituency. Integrating endangered species into ecosystem management will depend on many things, including a strong empirical knowledge base, ecological theories, computer models, adequate budgets, and technical expertise. But the ultimate focus must be on actual on-the-ground conditions. All else is merely preparatory.

Writing a prescription for success may be fairly simple; turning that vision into reality will undoubtedly be difficult. At this time, at least four elements are essential to a successful ecosystem approach to conserving endangered species:

- Clear, consistent policies for biological diversity that transcend agency jurisdictions and bring people together on common solutions to complex problems
- Integrative research, technology development, and application focused on the key management question: How do we conserve endangered species through management of ecosystems that serve many other land uses?
- Innovative marketing to build awareness, create a more informed constituency for endangered species and biodiversity goals, and secure the resources needed to meet those goals
- Prioritized actions and monitoring to meet specific objectives for genetic resources, species populations, biological communities, and ecological processes through carefully designed adaptive resource management strategies

Note

1. I use the term *endangered species* throughout this essay to refer to species formally listed or being considered for listing by federal or state government. The concern here is not so much for legal classification as for uncertainty in future existence and how to address it. Biological diversity is the variety of life and its processes in an area. An ecosystem approach to conservation views land as a community of soils, waters, and biota that must retain its compositional, structural, and functional integrity (Franklin, 1988) to sustain both its biological diversity and yields of desired resources.

References

Cairns, J., Jr. 1989. "Restoring Damaged Ecosystems: Is Predisturbance Condition a Viable Option?" *Environmental Professional* 11:152–159.

Franklin, J. F. 1988. "Structural and Functional Diversity in Temperate Forests. Pp. 166–176 in E. O. Wilson (ed.), *Biodiversity*. Washington D.C.: National Academy Press.

Harris, L. D. 1984. *The Fragmented Forest: Island Biogeography Theory and the Preservation of Biotic Diversity*. Chicago: University of Chicago Press.

Holling, C. S. (ed.). 1978. *Adaptive Environmental Assessment and Management*. Chichester, U.K.: Wiley and Sons.

Landres, P. B., J. Verner, and J. W. Thomas. 1988. "Ecological Uses of Vertebrate Indicator Species: A Critique." *Conservation Biology* 2(4):316–328.

Leopold, A. 1949. *A Sand County Almanac*. New York: Oxford University Press.

Lovejoy, T. E., and D. C. Oren. 1981. "The Minimum Critical Size of Ecosystems." In R. L. Burgess and D. M. Sharpe (eds.), *Forest Island Dynamics in Man-Dominated Landscapes*. New York: Springer-Verlag.

Myers, N. 1988. "Tropical Forests and Their Species: Going, Going . . . ?" Pp. 28–36 in E. O. Wilson (ed.), *Biodiversity*. Washington D.C.: National Academy Press.

Naess, A. 1986. "Intrinsic Value: Will the Defenders of Nature Please Rise?" In M. E. Soulé (ed.), *Conservation Biology: The Science of Scarcity and Diversity*. Sunderland, Mass.: Sinauer Associates.

Newmark, W. D. 1987. "A Land-Bridge Island Perspective on Mammalian Extinctions in Western North American Parks." *Nature* 325:430–432.

Norton, B. G. 1988. "The Constancy of Leopold's Land Ethic." *Conservation Biology* 2(1):93–102.

Pickett, S.T.A., and J. N. Thompson. 1978. "Patch Dynamics and the Design of Nature Reserves." *Biological Conservation* 13:27–37.

Quinn, J. F., C. van Riper III, and H. Salwasser. In press. "Mammalian Extinctions from National Parks in the Western United States." *Ecology*.

Salwasser, H. 1988a. "Editorial." *Conservation Biology* 1(4):275–277.

Salwasser, H. 1988b. "Managing Ecosystems for Viable Populations of Vertebrates: A Focus for Biodiversity." In J. K. Agee and D. R. Johnson (eds.), *Ecosystem Management of Parks and Wilderness*. Seattle: University of Washington Press.

Salwasser, H., S. P. Mealey, and K. Johnson. 1984. "Wildlife Population Viability: A Question of Risk." *Transactions of the North American Wildlife and Natural Resources Conference* 49:421–439.

Salwasser, H., C. M. Schonewald-Cox, and R. Baker. 1987. "The Role of Interagency Cooperation in Managing for Viable Populations. Pp. 159–173 in M. E. Soulé (ed.), *Viable Populations for Conservation*. New York: Cambridge University Press.

Salwasser, H., and others. 1989. "A Marketing Approach to Wildlife and Fisheries Program Management." *Transactions of the North American Wildlife and Natural Resources Conference* 54:261–270.

Schonewald-Cox, C. M., and J. W. Bayless. In press. "The Boundary Approach: A Geographic Analysis of Design and Conservation of Nature Reserves." *Biological Conservation.*

Servheen, C. 1984. "The Status of the Grizzly Bear and the Interagency Grizzly Bear Recovery Effort." *Proceedings of the Western Association of Fish and Game Commissioners.* Victoria, B.C.: Western Association of Fish and Game Commissioners.

Thomas, J. W. (ed.). 1979. *Wildlife Habitats in Managed Forests: The Blue Mountains of Oregon and Washington.* USDA Forest Service Agriculture Handbook No. 553. Washington, D.C.: U.S. Department of Agriculture.

Verner, J. 1984. "The Guild Concept Applied to Management of Bird Populations." *Environmental Management* 8:1–14.

Walters, C. 1986. *Adaptive Management of Renewable Resources.* New York: Macmillan.

Wilson, E. O. (ed.). 1988. *Biodiversity.* Washington, D.C.: National Academy Press.

World Resources Institute. 1989. *World Resources 1988–89.* Washington, D.C.: World Resources Institute.

Yaffee, S. L. 1982. *Prohibitive Policy: Implementing the Federal Endangered Species Act.* Cambridge: MIT Press.

COPING WITH IGNORANCE:
THE COARSE-FILTER STRATEGY FOR
MAINTAINING BIODIVERSITY

by

Malcolm L. Hunter, Jr.

O︎UR IGNORANCE of the natural world is enormous. Although we have been systematically cataloging the earth's inhabitants since the time of Linnaeus, it is widely believed that the vast majority of species have yet to be described. It has been estimated that the ratio of unknown to known species may be as much as 21:1—30 million unknown species versus 1.4 million species currently described (Wilson, 1988). This yardstick of our ignorance has been derived mainly by entomologists extrapolating from the frequency with which they find new insect species in tropical forests. It may be a conservative estimate.

Consider the relationship between the number of terrestrial animal species and their respective sizes depicted in Figure 6. It is not surprising that there are more species of small creatures than large ones. But why are there fewer species with a body length less than 5 millimeters than those with a body length of 5 to 10 millimeters? Is this difference real, or is it an artifact that reflects our meager attempts to distinguish very small species? What about cryptic or sibling species—species that look identical but are genetically isolated from one another and thus full-fledged species? The difficulties of identifying these species suggests that they may be far more common than we realize

FIGURE 6

RELATIVE ABUNDANCE OF TERRESTRIAL ANIMALS OF DIFFERENT SIZES

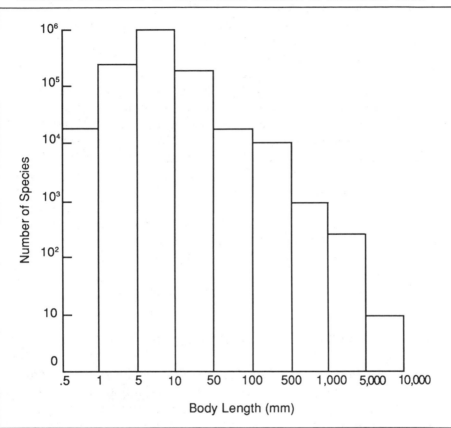

(Based on May, 1978, 1986)

(Bickham, 1983). Indeed, the whole idea of genetic isolation and species becomes rather nebulous for many plants and some of the simplest lifeforms, notably bacteria and viruses. In short, at this time we cannot even reliably estimate the number of species on earth to the nearest order of magnitude.

If we do not even know who the players are, our understanding of how well they are playing is far more deficient. It is only for a tiny portion of species that we have more than a vague notion of their abundance and distribution and ecological roles—and thus their conservation status. Looking beyond taxonomic classifications, our concern for biological diversity below the species level reveals an even greater dearth of knowledge about such issues as genetic diversity among populations. Similarly, our concern for the diversity represented in natural communities is often constrained by a lack of objective, meaningful ways to define them.

Against this backdrop of ignorance, the Endangered Species Act (ESA) may seem a pathetic actor, a token useful only to salve our conscience by saving a handful of the most charismatic species. And indeed, only a few species have been "recovered" since the first act was passed over two decades ago. Clearly this is an overly pessimistic view. True, the ESA is a powerful tool for conservationists in the United States that has positive ramifications far beyond the species directly affected. But as we come to realize all that we do not know, it has become increasingly obvious that the ESA, as it is currently implemented, is not enough to curb the tide of human-caused extinctions. For those species about which we know very little—again, the large majority of species—we must use another tool, one based on a higher level of biological organization, such as natural communities, to consider species in manageable groups.

COARSE FILTERS AND FINE FILTERS

The Nature Conservancy has a long tradition of protecting land of ecological importance. Thus it is not surprising that they were the first U.S. organization to articulate a strategy in which protecting natural communities complements saving endangered species (TNC, 1982; Noss, 1987). They explain their strategy with a metaphor: coarse filters versus fine filters. The basic idea behind the coarse-filter approach to maintaining biodiversity is to establish a set of reserves containing representative examples of all the various types of communities in a given area. If this array is reasonably complete, it is assumed that it will protect viable populations of most species. For the remaining species— those that fall through the pores of a coarse filter—a series of fine filters are needed. Fine filters are individually tailored conservation plans for those species that require them.

The federal endangered species program exemplifies the fine-filter approach. It sets up a system whereby individual species (or subspecies) are first identified as threatened or endangered. Then a recovery plan is put into action for each of these species—at least in theory. Habitat acquisition and restoration, captive breeding, reintroduction programs, and a variety of other activities are undertaken with the ultimate goal of restoring the species to a point at which it no longer requires protection under the act. Experience has shown, however, that it is difficult, if not impossible, to keep up with the growing number of species in need of assistance. In addition to the more than 1,000 species currently listed as threatened or endangered, the U.S. Fish and Wildlife Service has compiled a list of nearly 3,000 candidates for official listing.

The coarse-filter/fine-filter model may offer a means to tackle the seemingly endless backlog and cope with information gaps. The limitations of the fine-filter model can be mitigated by allying it with a coarse-filter strategy and vice versa. In the best of all possible worlds, no species would fall through the pores of this two-tiered plan. The following discussion will focus on the advantages and limitations of the coarse-filter strategy as a complement to the traditional fine-filter approach to species conservation. Improvements to the coarse-filter strategy are also discussed, along with prospects of incorporating the model into a national policy.

Advantages of the Coarse-Filter Strategy

The advantages of the coarse-filter strategy are rather straightforward. First, it can operate with relatively little information. If one has enough knowledge to define communities meaningfully, the communities can be black boxes with regard to their internal structure. This advantage is particularly important because, even outside the tropics, there is never enough information about microorganisms and seldom enough about invertebrates and nonvascular plants. Second, a coarse-filter strategy is efficient. It is efficient simply because it is cheaper to manage ecosystem units than to manage myriad endangered species scattered over the landscape. It is also efficient because the coarse-filter approach will maintain viable populations of nonendangered species, as well, minimizing the possibility that they will require an expensive fine-filter approach in the future. Finally, a coarse-filter strategy maintains the integrity of whole ecological systems, thereby sustaining the ecological and evolutionary context in which organisms exist.

Limitations of the Coarse-Filter Strategy

COMPREHENSIVENESS

The most obvious limitation of the coarse-filter strategy is that it cannot be comprehensive. Some species will fall outside its purview and require fine-filter strategies. Species under extreme threat from overexploitation, pollution, and similar exigencies need fine-filter management—the most egregious current example is probably the poaching of black rhinos. Threats need not be human in origin, however. Brood parasitism by cowbirds is a threat to Kirtland's warblers that has little to do with protecting the communities in which the warblers live.

Species that occur only at a few, scattered locations cannot be covered by the coarse-filter approach without stretching the concept. Protecting an array of representative communities in Texas would be unlikely to save the Texas wild rice, for example, a species confined to less than a hectare along the San Marcos River—unless one defined a Texas wild rice community type. Defining a community around a single rare species is possible, but it does contradict the basic idea of a coarse filter.

Species that occur at very low population densities (such as the large carnivores) do not fit into the coarse-filter approach well, either, because they are often habitat generalists not particularly tied to any type of community. Moreover, protecting viable populations of such species involves very large areas. Herein lies a second limitation of the coarse-filter strategy: scale.

SCALE

Ecologists often think of a natural community as a set of interacting populations that occupies a defined space—for example, one community ends and another begins at the lakeshore or the birch stand/spruce stand edge, even though significant interactions may take place across this boundary. Communities defined in this manner are usually too small to maintain viable populations of all their constituent species if they become isolated islands of natural habitat in a sea of communities shaped by humans. This approach also ignores the fact that many communities do not have sharp boundaries but grade into other communities across broad transition zones.

Avoiding these problems requires operating at a landscape scale (hundreds and thousands of square kilometers) that greatly exceeds the scale at which ecologists typically think of communities (tens, hundreds, and thousands of hectares). (See Noss, 1983, 1987; Baker, 1989.) One could define communities at a larger scale using boundaries such as watersheds, but this would distort the established concept of a community.

COMMUNITY DOMINANTS

Another limitation arises because communities are typically defined by their dominant species, especially the conspicuous plants that comprise most of their biomass. One hears of oak–pine forests, cypress swamps, and cattail marshes, not lily forests and heron marshes. By definition, plants that are community dominants are successful, abundant species, and often they have a broad tolerance for different environmental conditions. If their tolerance is much broader than that of most species, it can be misleading to try to predict the distribution of a whole assemblage of species (that is, a community) from the distribution of a few dominants.

For example, forests dominated by red spruce are a common and readily recognized community in the northeastern United States (Society of American Foresters, 1967); however, their species composition differs among different environments (Hunter and others, 1988). In high-elevation red spruce communities, mountain ash, mountain paper birch, and mountain maple are common trees and gray-cheeked thrushes and blackpoll warblers are likely to be present. Poorly drained, lowland red spruce stands are more likely to have Swainson's thrushes, bay-breasted warblers, red maple, and northern white cedar. On coastal islands, the red spruce stands most closely resemble montane stands, except they have Swainson's thrushes instead of gray-cheeked thrushes and lack mountain maples. Red maple swamps in Vermont provide a second example; the set of co-dominant trees and understory plants depends strongly on the environment (riverine floodplain, lakeshore, isolated basin) in which the swamp is situated (Golet and others, in prep.).

In theory, communities cculd be defined by less conspicuous species (often called indicator species) that have distributions typical of many other species. In practice, the concordance in distribution among different species is often weak unless they are tightly coevolved, such as parasite and host. There are many examples of coevolution leading to parallel distributions, but they represent a small minority of potential relationships for reasons explained in the next section.

PREDICTABLE COMMUNITIES

An unstated assumption of the coarse-filter strategy is that communities are reasonably stable, predictable assemblages of species. If they are not, it becomes difficult to devise a meaningful classification system as a basis for protecting a representative array. This assumption is easily challenged, however, when communities are viewed over extended periods (Hunter and others, 1988). The paleoecological literature reveals that the composition of communities is constantly changing as species shift their geographic ranges in individualistic response to climate change.

This phenomenon has been demonstrated by George Jacobson and colleagues (1987), who used pollen stratigraphy from lake sediment cores throughout eastern North America to map lines of equal pollen abundance for eight species and genera of plants during the last 18,000 years. To take one example, they showed that beech and hemlock, which are currently dominants in a widely recognized community type of northeastern North America (northern hardwoods), had divergent ranges just 12,000 years ago. At that time, beeches were primarily found in the lower Mississippi valley and hemlocks in the central Appalachians. Small mammals provide a converse example. Currently black-tailed prairie dogs, northern bog lemmings, and eastern chipmunks have different ranges and are members of three very different communities—prairies, boreal forests, and temperate deciduous forests, respectively. Fossil evidence indicates that they all coexisted in the same area, western Iowa, 23,200 years ago (Graham, 1986).

Worrying about the fact that communities are just transitory collections of species when viewed from a millennial perspective may seem excessively long-visioned in light of the immediacy of current threats to biological diversity. Yet we may soon be dealing with issues of biotic response to climate change in a much shorter time frame. Not only does the greenhouse effect threaten to accelerate climate change (National Research Council, 1988), but there is evidence that even historical climate change may have been far more rapid than usually thought. Data from Greenland indicate that 10,700 years ago there was a period of just twenty to fifty years during which there was a local 7°C warming and a 50 percent increase in precipitation (Dansgaard and others, 1989).

Disturbance and succession also confound classification systems for many communities, especially forests, because they can effect such significant changes in species composition and these changes are not always predictable. Ecologists often circumvent this problem (at least on paper) by using just climax communities for their classifications—sometimes called "potential vegetation" to avoid the controversy around the concept of climax (Küchler, 1964).

Insufficient information

Although the coarse-filter strategy is predicated, in large part, on the idea that biodiversity can be maintained despite a dearth of information, some information is necessary. Ideally one would discern key biological boundaries from accurate maps of species distributions that could be correlated with one another and compared to soil, topography, and climate. These maps are often not available, however, at least at the level of accuracy one would desire. In the absence of such detailed knowledge there is still a role for the coarse-filter strategy. For example, an analysis of biodiversity in Nepal used the coarse-filter approach with just one, easily measured, environmental parameter: altitude (Hunter and Yonzon, in prep.). Altitudinal profiles of the distribution of Nepali birds, mammals, and parks made it very clear that there was a significant shortage of protected areas at intermediate altitudes even though these areas had a rich fauna. As always, however, conservation biology is a crisis discipline in a rapidly changing world (Soulé, 1985). Thus we must be willing to work with whatever information is available.

IMPROVING THE COARSE-FILTER STRATEGY

The most obvious limitations of the coarse-filter approach are easily circumvented, at least conceptually, by allying it with its complement, the fine-filter strategy, to form a comprehensive program. Making the coarse-filter component effective, however, would require several modifications.

Ecosystems

A coarse-filter strategy should not be based solely on communities. It should consider both communities and the physical environments they occupy—in other words, it should be an "ecosystem-based" strategy. Ecologists usually delineate ecosystems by their plant communities. (Marine ecosystems are a common exception.) However, this emphasis on species distributions needs to be reexamined in light of paleoecological research that has shown physical environments to be the arenas of biodiversity and evolution while communities are just the temporary, constantly changing, occupants of these arenas. Indeed, for this reason one could argue that physical environments may be a better basis for a coarse-filter strategy than communities (Hunter and others, 1988).

It may seem counterintuitive to protect biodiversity by focusing on the physical components of ecosystems, such as climate and soil, rather than on assemblages of plants and animals. Obviously, biodiversity is asters and ants, not nitrogen, photons, and water. The main reason for concentrating on physical factors is that they are likely to provide a more broadly relevant basis for classifying ecosystems. Key environmental factors such as temperature, water budget, and nutrient availability affect all organisms. In contrast, the distribution of any given species may be of little direct importance to most of the other biota in the ecosystem; it may not even show much concordance with other distributions. In the red spruce forest, for example, the distribution of red spruce is a poor predictor of the distributions of other species even at one point in time. Over an extended period, it would be particularly inadequate.

Of course it would make little sense to define ecosystems arbitrarily by their physical factors—for example, ecosystem type x has an annual precipitation of 75–100 cm, a mean annual biotemperature of 10–12°C, a soil pH of 6–7, and concentrations of cobalt, selenium, and vanadium above 1 ppm. One must consider which factors critically affect the distribution of organisms. Ideally one would also search for points along environmental gradients (1 ppm? 10 ppm?) that may be key biological thresholds. The relationships between physical environments and biotic distributions are widely understood in a general sense, but the depth of understanding is rather poor. Further work by ecological physiologists is much needed. Such work might elucidate the physiological processes that shape distributions and thus identify some critical environmental thresholds.

Another advantage of using the distributions of key physical factors to classify and identify ecosystems is that they are often reasonably well documented—either directly in climate and soil maps or indirectly in maps of bedrock geology and topography. In some ecosystems it may be more efficient to use indicator species whose distributions manifest important ecological patterns—for example, in wetland ecosystems many plants are quite sensitive to the hydologic regime. However, unless the indicator species are dominant enough to form vegetation identifiable by remote sensing, it may not be easy to determine their distributions on a large scale.

Finally, an environmentally based system of ecosystem classification would cope well with the complexities created by patterns of disturbance and succession. In some contexts, it is important to distinguish between a mature spruce forest community and a patch of raspberries that will become a mature spruce forest in a few decades. For many long-term conservation programs, the distinction is trivial. It is more important to know that the type of physical environment which supports both these communities will be protected as an arena for an ever-changing continuum of populations.

By emphasizing some of the advantages of focusing on physical features, I do not mean to diminish the importance of the biological components of ecosystems. Obviously biodiversity resides with the biota, and often ecosystems will be most readily classified and identified by their constituent species. As explained above, however, there are significant advantages to considering environmental factors that have often been overlooked in favor of conspicuous species.

Biophysical Regions

To further enhance the efficacy of the coarse-filter strategy, it is useful to think of the regional context in which ecosystems exist. Red spruce ecosystems in Tennessee, New York, Maine, and Prince Edward Island are likely to be quite different, but distinguishing between lowland and montane environments will not necessarily be helpful in making these differentiations. One should also consider the broad, regional patterns of physical features that shape the geographic distributions of species and can thus define biophysical regions. Climate is the primary environmental factor that defines a biophysical region, while gross patterns of soils and surface geology may serve as secondary distinguishing factors.

In Maine, climate, soils, and surface geology were used to divide the state into fifteen physical regions averaging 5,352 square kilometers each. Boundaries were then adjusted by cross-checking them against the geographic ranges of woody plant species (McMahon, 1990). Because biological data were used to corroborate the importance of the physical regions, these units can be considered biophysical regions. In the next phase of the program, each of these regions will be inventoried to determine what types of ecosystems occur there and which types have been adequately protected. Finally, an effort will be made to fill in the gaps by locating and ensuring the protection of examples of each type of ecosystem identified. This gap analysis approach to implementing the coarse-filter strategy has been advocated frequently. (See Scott and colleagues in this volume; Burley, 1988; Crumpacker and others, 1988; DeVelice and others, 1988.) However, most analyses of this type operate only at a regional scale, rather than at both a local and regional scale. In addition, such studies tend to use communities or vegetation types rather than ecosystems (although they may be called ecosystems, as by Crumpacker and others, 1988).

Scale

Using large-scale environmental parameters to define biophysical regions helps to refine the classification of ecosystems. It does nothing, however, to mitigate the facts that a single ecosystem is typically too small to maintain viable populations of all its species, especially larger animals, and that many species require habitat with juxtaposed ecosystems—such as a forest beside a lake or a late successional ecosystem next to an early successional ecosystem. Since these issues have been discussed at length elsewhere, only summary recommendations will be presented here: Choose large examples of ecosystems; select groups of ecosystems in assemblages that form natural landscape units such as watersheds and mountain ranges (Noss 1987 recommends ways to classify, inventory, evaluate, and manage such ecosystem mosaics); and connect landscape units with habitat corridors. The simplicity with which these recommendations are presented belies considerable complexity, for limited resources nearly always make it impossible to follow an ideal plan.

A NATIONAL POLICY

Adding an Endangered Ecosystems Act to complement the Endangered Species Act would be a considerable task because ecosystems are not as tangible as species. They have no big, brown eyes to endear themselves to the public; they are not even very tangible to ecologists, who have no precise way to define them. Nevertheless, such a measure is an achievable goal. Witness the Emergency Wetlands Resources Act of 1986 and the North American Wetlands Conservation Act of 1989—essentially endangered ecosystem acts for a particular set of ecosystems. (See Noss and Harris, 1986; Hunt, 1989; and Noss' essay in this volume.)

What form should such an act take? Funding for public acquisition of ecosystem types that are rare, extensively threatened by degradation, or underrepresented among currently protected areas is the most obvious component. This would be a straightforward program that could soon produce significant results. Such a program might be sufficient if one were only talking about a modest series of purchases: a relict dune system along the Great Lakes, a serpentine formation in the Sierra Nevada, a calcareous fen in the Northeast. However, the vulnerability of small ecosystems dictates that large areas be protected to encompass clusters of ecosystems that are connected with other clusters by ecological corridors. And to accomplish this end may require more funding than is likely to be forthcoming.

Ecosystems could also be protected by regulations prohibiting any activity that would significantly degrade them. Elsewhere in this volume Anne and Paul Ehrlich propose severe limitations on development (including agriculture and forestry) of all ecosystems still in a relatively pristine state: virgin forests, unplowed prairies, most wetlands, and other ecosystems where the only significant human intrusions are global in scope, such as air pollution. In most places such ecosystems are so rare that protecting them would have virtually no impact on the local economy. If protection places an undue burden on the landowners, taxpayers should be willing to pay for maintenance of a societal value by providing at least property tax relief to the landowners.

Moreover, ecosystems that have been subject to intensive use such as logging and grazing, but still remain essentially natural ecosystems, should not be overlooked. If sensitively managed, they can provide an income to their owners while contributing significantly to the maintenance of biological diversity (Hunter, 1990), particularly by serving as buffer zones around completely protected areas (Noss and Harris, 1986). Again, it may be necessary to provide government subsidies to the land-owners to compensate them for the cost of managing their land for diversity. (For an analysis of the relative merits of having landowners, taxpayers, consumers, or outdoor recreationists pay these costs see Hunter, 1990, chap. 15.) In many cases, the restoration of degraded ecosystems should be pursued as well; indeed, in some instances (such as prairies and eastern forests), this is the only option generally available.

Maintaining the earth's biological diversity in the face of an overwhelming range of threats is a daunting task, especially because the elements of biodiversity are truly innumerable. The coarse-filter strategy offers a conceptual tool for organizing much of this diversity into manageable units—the assemblages of populations called communities. It is based on the assumption that by protecting a representative array of communities, the majority of species can be maintained. Only a minority of species that fall through the pores of this coarse filter require individual management—the fine-filter approach.

Converting the coarse-filter strategy into a practical tool involves dealing with two of its limitations. First, the species composition of communities is not very predictable, especially over a period of time. Thus it is necessary to think in terms of both communities and their physical environments (in other words, ecosystems), for physical environments are the arenas of biodiversity whereas communities are just the temporary occu-pants. The second major problem is one of scale. Because ecosys-tems are usually too small to maintain viable populations of all their species, they must be protected in clusters that form large landscape units. Using the tool will require not only the acquisi-tion of key ecosystems but also new laws to protect pristine ecosystems, restore degraded ones, and foster sensitive use of ecosystems managed for their natural resources.

Note

I thank Aram Calhoun, Kathryn Kohm, Janet McMahon, Reed Noss, and Michael Soulé for their helpful suggestions.

References

Baker, W. L. 1989. "Landscape Ecology and Nature Reserve Design in the Boundary Water Canoe Area, Minnesota." *Ecology* 70:23–35.

Bickham, J. W. 1983. "Sibling Species." Pp. 96–106 in C. M. Schonewald-Cox and others (eds.), *Genetics and Conservation*. Menlo Park, Calif.: Benjamin/Cummings.

Burley, F. W. 1988. "Monitoring Biological Diversity for Setting Priorities in Conservation." Pp. 227–230 in E. O. Wilson (ed.), *Biodiversity*. Washington, D.C.: National Academy Press.

Crumpacker, D. W., S. W. Hodge, D. Friedly, and W. P. Gregg. 1988. "A Preliminary Assessment of the Status of Major Terrestrial and Wetland Ecosystems on Federal and Indian Lands in the United States." *Conservation Biology* 2:103–115.

Dansgaard, W., J.W.C. White, and S. J. Johnsen. 1989. "The Abrupt Termination of the Younger Dryas Climate Event." *Nature* 339:532–534.

DeVelice, R. L., J. W. DeVelice, and G. N. Park. 1988. "Gradient Analysis in Nature Reserve Design: A New Zealand Example." *Conservation Biology* 2:206–217.

Dial, K. P., and J. M. Marzluff. 1988. "Are the Smallest Organisms the Most Diverse?" *Ecology* 69:1620–1624.

Golet, F. C., A.J.K. Calhoun, W. DeRagon, D. Lowry, and A. J. Gold. In prep. "The Ecology of Red Maple Swamps in the Glaciated Northeastern United States: A Community Profile." U.S. Fish and Wildlife Service.

Graham, R. W. 1986. "Response of Mammalian Communities to Environmental Changes During the Late Quaternary." Pp. 300–313 in J. Diamond and T. J. Case (eds.), *Community Ecology*. New York: Harper & Row.

Hunt, C. E. 1989. "Creating an Endangered Ecosystems Act." *Endangered Species Update* 6(3–4):1–5.

Hunter, M. L., Jr. 1990. *Wildlife, Forests, and Forestry: Principles of Managing Forests for Biological Diversity*. Englewood Cliffs, N.J.: Prentice-Hall.

Hunter, M. L., Jr., and P. B. Yonzon. In prep. "Altitudinal Distribution of Birds, Mammals, Forests, Parks, and People in Nepal."

Hunter, M. L., Jr., G. L. Jacobson, Jr., and T. Webb III. 1988. "Paleoecology and the Coarse-Filter Approach to Maintaining Biological Diversity." *Conservation Biology* 2:375–385.

Jacobson, G. L., Jr., T. Webb III, and E. C. Grimm. 1987. "Patterns and Rates of Vegetation Change During the Deglaciation of Eastern North America. Pp. 277–288 in W. F. Ruddiman and H. E. Wright, Jr. (eds.), *North America and Adjacent Oceans During the Last Deglaciation: The Geology of North America*, vol. K-3. Boulder: Geological Society of America.

Küchler, A. W. 1964. *Potential Natural Vegetation of the Conterminous United States*. New York: American Geographical Society.

May, R. M. 1978. "The Dynamics and Diversity of Insect Faunas." Pp. 188–204 in L. A. Mound and N. Waloff (eds.), *Diversity of Insect Faunas*. Oxford: Blackwell Scientific.

May, R. M. 1986. "The Search for Patterns in the Balance of Nature: Advances and Retreats." *Ecology* 67:1115–1126.

McMahon, J. 1990. "The Biophysical Regions of Maine." Master's thesis, University of Maine, Orono.

National Research Council. 1988. *Toward an Understanding of Global Change*. Washington, D.C.: National Academy Press.

Noss, R. F. 1983. "A Regional Landscape Approach to Maintain Diversity." *BioScience* 33:700–706.

Noss, R. F. 1987. "From Plant Communities to Landscapes in Conservative Inventories: A Look at The Nature Conservancy (USA)." *Biological Conservation* 41:11–37.

Noss, R. F., and L. D. Harris. 1986. "Nodes, Networks, and MUM's: Preserving Diversity at All Scales." *Environmental Management* 10:299–309.

Society of American Foresters. 1967. *Forest Cover Types of North America*. Washington, D.C.: Society of American Foresters.

Soulé, M. E. 1985. "What Is Conservation Biology?" *BioScience* 35:727–734.

The Nature Conservancy (TNC). 1982. *Natural Heritage Program Operations Manual*. Arlington, Va.: The Nature Conservancy.

Wilson, E. O. 1988. "The Current State of Biological Diversity." Pp. 3–18 in E. O. Wilson (ed.), *Biodiversity*. Washington, D.C.: National Academy Press.

GAP ANALYSIS OF SPECIES RICHNESS AND VEGETATION COVER: AN INTEGRATED BIODIVERSITY CONSERVATION STRATEGY

by

J. Michael Scott, Blair Csuti, Kent Smith, J. E. Estes, and Steve Caicco

Extinction has become a major issue attracting worldwide attention. While the list of endangered, threatened, and sensitive species in North America is already depressingly long, it continues to grow. Many of the species that will be added to the list are ones that we now consider uncommon or locally common. Today, as we reflect on the years since the passage of the Endangered Species Act (ESA), we see a record of extinction unforeseen by those responsible for the act.

Currently the U.S. Fish and Wildlife Service lists over 1,000 taxa as endangered or threatened. An additional 3,000 or so species are identified as either Category 1 or 2 species. These include plants and animals for which sufficient information exists to propose listing (Category 1) and those believed to be threatened or endangered but for which more data are required to meet the legal requirements for listing (Category 2). While these figures may seem staggering, they do not begin to tell the whole story. A report recently commissioned by the California Nature Conservancy, for example, estimated that 220 animals,

600 plants, and 200 natural communities in California alone may be threatened with severe reduction or even extinction (Jones and Stokes Associates, 1987). Norman Myers (1979) has projected 5,000 extinctions per year, worldwide, in the near future. Yet in the years since the ESA was passed, only 229 recovery plans have been approved. Many of these elaborate and costly plans have been written for species numbering less than a hundred individuals.

We have learned much since the passage of the Endangered Species Act. Our greatest lesson, however, may be that despite the act's best intentions species continue to go extinct as a result of human activity and the list of threatened and endangered species continues to grow. Our current programs to address the extinction problem have become essentially efforts to document the loss of species through the listing process. Our emphasis on saving critically endangered species, one at a time, has too frequently resulted in crisis management for individual plants and animals—many of which have little hope of ever recovering.

A simple comparison illustrates the dilemma we face. When we compare the burgeoning list of threatened and endangered species with the tremendous cost of recovering just one species, only one conclusion can be reached: Regardless of how we feel about the value of these species, social, economic, and biological realities preclude saving all of them. From strictly a fiscal perspective, to make significant advances toward recovery of many, if not most, taxa would take millions of dollars more than we can ever reasonably expect to be allocated. There should be little wonder, then, over the obvious frustration that resource professionals and concerned citizens feel when attempting to deal with endangered species issues. As the situation continues to worsen, frustrations and confrontations increase to the point where a sense of hopelessness prevails. What compounds the situation is that laws, policies, and, to a great extent, public opinion tend to focus our financial and intellectual resources on individual species.

We believe that a significant change in focus in addressing the endangered species problem is warranted and possible—and that we do have a choice. We can continue to focus on critically endangered species one at a time and allow individual recovery efforts to be diluted each time a new plant or animal is added to the list. Or we can proceed in a more positive direction by shifting some of our efforts to a more broad-based ecosystems approach aimed at *preventing* species from becoming endangered. By integrating new and existing conservation strategies, we can begin to advance from our traditional single-species management approach toward a land-based philosophy with the goal of conserving overall biodiversity. In this essay, we describe such an approach using "gap analysis" as part of an integrated conservation strategy for the preservation of diversity. Certainly it is not the only approach. But it is one that provides a solid basis, given current technology, for addressing a complex and critical conservation need.

Even game species could be reduced to nonharvestable numbers, or even to the point of endangerment, in the next century. These reductions will result largely from predictable but avoidable loss of wildlife habitat through future patterns of land use and management. Thus we face a greater challenge than recovering individual species: ensuring the integrity of existing natural communities and ecosystems, thereby minimizing the number of critically endangered species in the future. In terms of protecting biodiversity, the bottom line is not the number of species we save from the threshold of extinction in the next decade but the richness of our flora and fauna in a hundred years. Only a small fraction of earth's biological diversity can be kept in cryogenic arks or in "protective custody." If biodiversity is to be saved, our focus must be on saving functioning ecosystems.

Programs for species on the brink of extinction should not be abandoned. Rather, we need a more balanced conservation strategy to supplement these recovery programs with actions that will protect the vast majority of earth's species—those that are not considered rare or endangered but whose habitat nevertheless is shrinking. It is easier and more cost-effective to protect intact, functioning ecosystems with their myriad species than to initiate emergency room measures for one endangered species after another or to wait until common species become endangered before acting to protect them. Such efforts are not only costly but have a low probability of success (Griffith and others, 1989).

The challenge, then, is to plan future patterns of growth and modifications in land use to ensure the survival of most of the remaining biodiversity. This goal will not be reached by rescuing specific endangered species but by keeping enough of the living world around to supply us with, among other things, disease-resistant strains of crop plants, new medicines for new diseases, and functioning watersheds that supply water for drinking and irrigation. We offer gap analysis of existing vegetation types and species richness as a new approach that could place resource managers in an active rather than reactive position.

THE CONCEPT

The most obvious way to determine the extent of biodiversity being protected is to determine the species and communities that occur in protected areas. Native species that have adapted to the human environment are excluded from this analysis, as are nonnative species. All unprotected species and communities represent "gaps" in the conservation safety net (Burley, 1988). Because most species are restricted to particular habitats and biogeographic regions, an analysis of existing preserves in relation to vegetation types and to centers of high species richness will identify gaps in the present network of preserves. If collectively managed to maintain their biodiversity, these unprotected areas of high species richness and unprotected vegetation types would capture the majority of continental biological diversity (Scott and others, 1987).

There are a number of approaches to gap analysis, each with its strengths and weaknesses. The most widely used indirect assessment of the distribution of biodiversity relies on vegetation types as indicators of specific communities of plants and animals. (See Diamond, 1986; Huntley, 1988; Crumpacker and others, 1988; Backus and others, 1988.) Although an inventory of protected plant communities is clearly an important component of gap analysis, there are several critical shortcomings that can only be addressed by direct measurements of the distribution of biodiversity. The most pressing of these problems concern spatial scale and validity.

Unlike species, however, vegetation types or associations are artificial and often ambiguous entities. No universally accepted system of vegetation classification exists. In fact, classifications of natural communities are usually specific to states or regions and are unique in their purpose and structure. Often this situation is compounded because numerous classifications have been developed for a single state or region. At the continental scale, IUCN and UNESCO reviewed the protection of twenty-four biogeographic provinces in North America (Udvardy, 1984), while Crumpacker and colleagues (1988) investigated the representation of 135 Küchler "potential natural vegetation" (PNV) types on federal and Indian lands in the United States alone. Within the state of California, various classification schemes have identified from fifteen to seventy plant communities (Barbour and Major, 1988)—and many systems describe subdivisions of even those communities (Cheatham and Haller, 1975; Paysen and others, 1980). The California Department of Fish and Game in cooperation with the California Nature Conservancy has identified 375 natural communities in California and is working on establishing protection priorities for them (Holland, 1987). There is no agreement among conservation biologists on which levels in these classification hierarchies biodiversity is best represented or can best be protected. For many regions, the process of classifying vegetation is incomplete or disputed. This makes it virtually impossible to reach consensus on how conservation priorities should be set. Further, as Reed Noss (1987a, p. 11) points out, "some important ecological functions of heterogeneous landscapes . . . are not necessarily protected by conservation strategies that focus on separate, homogeneous community types."

The question of appropriate scale is not the only problem with a vegetation-based gap analysis. Plant communities may exist in various stages of disturbance; this is of utmost importance to their conservation priority. Crumpacker and colleagues (1988) describe the difficulty of using Küchler PNV types as a basis for predicting current cover and land-use practices. They note, for example, that "forested types in very early successional stages often may be only slightly more valuable for conservation purposes than . . . croplands or pasturelands" (p. 112). It is questionable whether the prairie communities of the Great Plains would ever revert to their Küchler PNV types, even if agricultural activity were to cease immediately. This creates significant problems in interpretation.

Verification of vegetation identity and status is a critical but rarely practiced component of vegetation classification and mapping. The use of remote-sensing data, especially those generated by high-altitude aerial photography and satellite imagery, may improve this situation in the future (Mayer, 1984; Scott and others, 1987). Nevertheless, some type of validation is essential to an accurate assessment of the range of variability in species composition and structure within a mapping unit.

We believe that gap analysis based on cover type alone provides an incomplete picture of biodiversity. A better approach would be to include information on species richness for those taxa for which biologically defensible data on distribution are available. Adequate distribution information exists for most species of vertebrates and for some plants and subgroups of invertebrates (primarily butterflies). Although vertebrates make up only a small fraction of all named species, perhaps 3 percent (Wilson, 1988), their mobility and habitat specificity make them useful indicators of overall biodiversity.

Most floras provide only the broadest description of the distribution of plant species. While intensive fieldwork can lead to a species-by-species assessment of protection at the local level (Margules and others, 1988), the botanical component of biodiversity is usually approached from the vegetation level. A gap analysis based on individual plant distributions would be possible only in those few regions with published plant atlases.

While the conservation of a few highly mobile vertebrates can provide an "umbrella" under which many other plants and animals are protected, the habitat requirements of vertebrates and invertebrates often differ enough that the conservation of one does not ensure that of the other (Murphy and Wilcox, 1986). Moreover, Robert M. Pyle (1982) points out several advantages that butterflies in particular have as regional biogeographical indicators: They have limited vagility; they are often host-specific to certain plants; their high reproductive potential minimizes changes in their distribution caused by human activity; and there are enough species to provide regional patterns but not so many as to overwhelm the analysis. Because of their host-plant specificity, they tend to "condense the vast amount of ecological information available in plants" (Pyle, 1982).

Thus we believe that a useful approach to developing a long-range strategy for preserving biodiversity is a multifaceted analysis of the gaps in the network of protected areas. Such an analysis would examine the distribution of several key elements of biodiversity (at scales of 1:250,000) relative to areas now under protective management or ownership. These would include the following:

- Vegetation types (actual rather than potential) at a scale of 1:250,000
- Terrestrial vertebrate distribution: centers of species richness for vertebrates grouped by taxa (nongame mammals, waterbirds, uncommon species, and so on) in each vegetation type and biogeographic province; centers of endemism; species-by-species protection status
- Terrestrial invertebrate (butterfly) distribution: centers of species richness in each biogeographic province; centers of endemism; species-by-species protection status
- Threatened, endangered, and sensitive species distribution
- Distribution of areas managed for the preservation of biodiversity, including public (federal, state, county) and private nature preserves, with an assessment of the degree of protection offered by present management

MATERIALS AND METHODS

The raw materials for a biogeographical study are maps depicting the distribution of species or communities of interest. Because many animal species prefer certain habitat types, a vegetation map is the highest priority for conducting a gap analysis for communities and delineating the ranges of terrestrial animals. The vegetation map should be as large in scale as practical. We suggest a scale of 1:250,000. This is large enough to show local detail but small enough to depict landscape organization and linkages. Many states do not have a single vegetation map but do have several maps covering different portions of the state. The best of these might be compiled into a single map.

Distribution maps of vertebrates and some invertebrates, particularly butterflies, are commonly prepared from the collective locations of records of occurrence. Many states have, or are developing, atlases depicting the distribution of species. If such distribution maps are unavailable, they can be prepared using known records of occurrence and can serve as the basis for mapping the probable limits of distribution. Range maps, however, can only assess the probability of encountering a species; they always contain patches of inappropriate habitat within the described range. The predictive ability of a range map increases to the degree that it is corrected by a large-scale habitat map and is verified by experts and fieldwork. This can be accomplished by conducting new field studies or by using previous floral and faunal surveys (not used to draw range maps) as independent checks of the validity of range maps. Vegetation maps can be verified using on-the-ground checks and up-to-date aerial photography.

Information on landownership is commonly mapped by state and federal agencies. The Bureau of Land Management (BLM) publishes maps of landownership and management status for many states. Private and state preserve areas can be individually plotted and added to these databases. Although some federal agencies are mandated to preserve biodiversity, different categories of federal ownership are managed for different purposes. Many national forests, for example, allow harvesting of stands of old-growth timber that, for practical purposes, can never be regenerated (Wilcove, 1988). BLM lands throughout the arid West are typically subject to intensive grazing with little monitoring to determine the status or trends in their biodiversity. A series of ranks indicating the level of protection should be applied to various categories of landownership and management. The Nature Conservancy assigns such categories: (1) total protection of native communities, (2) partial protection of native communities, and (3) no protection.

Recent advances in the development of "geographic information systems" (GIS) have made it possible to store and analyze multiple layers of geographic data on relatively inexpensive microcomputer systems. Once this information is stored, the computer can respond to search commands to identify gaps in the system of protected areas from a variety of perspectives. Comparison of these results makes it possible not only to determine which species or communities currently occur in areas managed for biodiversity but also to identify alternative strategies to achieve various levels of protection of other areas of high biodiversity (Scott and others, 1987).

Because many species are restricted to certain biogeographic provinces, it should be possible to identify sets of species whose distributions are correlated. Often several areas of richness will occur in a single province, offering alternative management strategies. Not surprisingly, local areas with considerable diversity of habitat or topography tend to have richer faunas and floras. The distribution of species within vegetation types can also be investigated. By directly measuring the distribution of an entire class of species (all mammals, all birds), the value of vegetation-based gap analyses can also be measured. This could be especially important for areas where species distributions are poorly understood (for example, most of the world's tropical moist forests) and where gap analysis based on vegetation maps may be the only practical approach to selecting preserves. The analysis of centers of high vertebrate and invertebrate (butterfly) richness may uncover important gaps in a conservation strategy based solely on one or the other group.

A multifaceted gap analysis at the state level might have the following stages:

1. Draft and digitize a map of vegetation type distribution.
2. Verify the vegetation map with fieldwork.
3. Draft and digitize vertebrate and invertebrate (butterfly) distribution maps.
4. Verify the animal distributions with fieldwork.
5. Input data on landownership status.
6. Generate a map depicting species richness.
7. Digitize current management areas and determine the level of protection.
8. Generate a map for special-interest species (such as threatened, endangered, and sensitive plants and animals, endemic taxa, or uncommon species found in fewer than three vegetation types).
9. Define and outline centers of species richness.
10. Compare lists of species represented in centers of richness to minimize the number of centers of richness (for example, analyze the data set for minimum redundancy) and maximize species and habitat representation.

11. Rank the centers of richness by their contribution to state, regional, and continental biodiversity.
12. Determine the current percentage of each area of species richness and vegetation type in protected areas.
13. Identify minimum and optimum areas required for protection of predetermined levels of statewide species richness.
14. Identify landscape corridors between areas of high species richness and natural vegetation types managed to provide self-sustaining natural communities.

The analysis would address the following questions:

- Are existing preserves located in areas of high species richness?
- Are threatened, endangered, or other species of special interest represented in protected areas?
- What is the ownership status of species-rich areas?
- What proportion of threatened, endangered, or sensitive species is protected in existing preserves?
- How will changes in land use affect the number of species not found in protected areas?
- What vegetation types are not found in protected areas?
- What species occur in protected areas? What species do not occur in protected areas? Which species are represented in the largest numbers of protected areas? Which species are represented in the fewest number of protected areas?
- Which unprotected areas should be given protection to include a viable population (Gilpin and Soulé, 1986) of each species in at least one preserve?
- Do adequate landscape corridors exist between areas of high species richness to provide for dispersal and interbreeding of populations?
- How will global warming affect the distribution of vegetation types and species richness within present and future preserves?

Once the gap analyses for a single state have been completed, the process could be extended to a regional and eventually a continental analysis. Data obtained during the verification process could be incorporated into databanks such as the Heritage program developed by The Nature Conservancy.

FIELD VERIFICATION OF SITES

A GIS-based gap analysis of the distribution of biodiversity and protected areas should be carried out at the finest level of resolution allowed by the data. The results will indicate the areas high in species richness in each state and the areas covered by vegetation types not represented in existing preserves. In many cases these areas will coincide. Because vegetation maps and species distribution maps can only predict the presence and condition of populations and communities in an area, field verification of key unprotected sites will be necessary.

Furthermore, each species and community has specific habitat and space requirements. Beyond confirming the presence of elements of diversity in areas with high protection priority, the minimum boundaries of proposed preserves or management units will have to be established in relation to species and community requirements. As Reed Noss (1987a) points out: "Animal species that require a combination of contiguous habitat types to meet life history needs may not be protected unless the inventory system explicitly recognizes habitat combinations in the landscape." Moreover, many communities exist in several successional stages, and the spatial and temporal aspects of patch dynamics define a minimum size for long-term persistence (Baker, 1989)—what Pickett and Thompson (1978) call the "minimum dynamic area." Finally, the population sizes and spatial diversity needed to provide particular plant and animal species with reasonable protection from extinction through genetic, demographic, or random events (Goodman, 1987) must be incorporated in the design of the preserve. In short, the description of recommended preserve or managed area boundaries becomes a key part of an integrated conservation strategy.

BIODIVERSITY OUTSIDE OF RESERVES

Readers will undoubtedly have a strong sense that we are presenting the concept of gap analysis as a mechanism for identifying key areas for inclusion into reserve systems. And, in fact, a comprehensive nature reserve network provides a critical foundation for maintaining biodiversity. But even the most progressive nature reserve systems protect only a small percentage of the land base that supports a nation's biodiversity. Worldwide only 1 or 2 percent of this land base is currently set aside in protected reserve status, and even these lands face insecurity as human designs place increasing demands upon them. The figure for protected areas in the United States is closer to 3 percent. Much unprotected land, however, remains in a more or less natural state and makes significant contributions to the overall maintenance of biodiversity. The potential for this contribution is especially high in the western United States, where about half the land is in public ownership.

The gap analysis approach, using a GIS database of vegetation types, species distributions, and habitat preferences, can also be used to monitor the impact of public land management practices on biodiversity. Patterns and intensity of resource exploitation (timber harvest, grazing, mining) can be tracked from a landscape perspective, monitoring habitat fragmentation and alteration. Many species survive in exploited habitats and will continue to be widely distributed over range and forest lands. Others—interior-forest-dwelling birds, for example—are sensitive to habitat patch size and disappear from excessively fragmented forests. A comprehensive GIS database can provide the information on the distribution of wildlife and habitat needed to ameliorate these conflicts, leading to a balance between biodiversity and resource use on unprotected lands. The amount, quality, and distribution of sensitive habitat types (such as riparian woodland in western rangelands) can also be monitored to establish levels of resource use compatible with the survival of species dependent on those habitat types.

Ultimately, we see biological and resource GIS databases being used to direct local, regional, and national land-use planning. As the maintenance of biodiversity becomes a national priority, the potential exists to manage entire regions as biosphere reserves. Noss (1987b) presents a strategy for landscape management that integrates human activities with biodiversity through a network of preserves, buffers, and corridors. Much of the world can still avoid the loss of biodiversity that accompanied the transformation of Europe into a landscape of cities, towns, pastures, and manicured forests. A network of ecosystem preserves, combined with intelligent exploitation of other wild-lands, offers this opportunity to North America. The inventory and monitoring of components of biodiversity through GIS technology is an avenue to that goal.

A MODEL PROGRAM

Although gap analyses for vegetation types have been undertaken for potential vegetation for California (Klubnikin, 1979) and the United States (Crumpacker and others, 1988), no attempt has been made to integrate the results with actual vegetation or patterns of species richness and endangered species locations to devise an integrated conservation strategy. Further, each of these studies represents a static picture of conservation needs and none offers generally accessible graphic displays of the results. Given that the opportunities for *in situ* preservation of biodiversity are rapidly vanishing in both the developed and underdeveloped nations, there is an urgent need to apply GIS technology to a multifaceted gap analysis of conservation needs. This approach is currently being tested in Idaho. Based on the results of this pilot project, a nationwide application could be completed before the year 2000.

The current emphasis on saving endangered species is diverting most of our conservation energy and resources to species on the brink of extinction. (See Scott and others, 1987; Hutto and others, 1987; and Csuti and others, 1987.) In the final analysis, however, the success of efforts to preserve biodiversity will be judged on the number of surviving species a hundred years from now, not on whether we save the California condor or black-footed ferret in the next decade. We need to act now to develop a strategy to accomplish this goal. Identifying gaps in the current network of preserves—the protection of which will save most of the remaining species and communities—is an efficient and cost-effective way to retain the maximum biodiversity in the minimum area.

Note

The Idaho Cooperative Fish and Wildlife Research Unit is funded and supported by the Idaho Department of Fish and Game, the University of Idaho, the U.S. Fish and Wildlife Service, and the Wildlife Management Institute. This is contribution number 395 from the University of Idaho, Forestry and Wildlife Resources Experiment Station. Funding for development of a gap analysis approach to saving biological diversity has been provided by the National Fish and Wildlife Foundation, the Idaho Department of Fish and Game, and the U.S. Fish and Wildlife Service. We thank David W. Crumpacker and Reid Goforth for thoughtful and helpful comments on early drafts of this essay.

References

Baker, W. L. 1989. "Landscape Ecology and Nature Reserve Design in the Boundary Waters Canoe Area, Minnesota." *Ecology* 70:23–35.

Barbour, M. G., and J. Major. 1988. *Terrestrial Vegetation of California*. 2nd ed. Sacramento: California Native Plant Society.

Blockstein, D. E. 1988. "U.S. Legislative Progress Toward Conserving Biological Diversity." *Conservation Biology* 2:311–313.

Burley, F. W. 1988. "Monitoring Biological Diversity for Setting Priorities in Conservation." Pp. 227–230 in E. O. Wilson (ed.), *Biodiversity*. Washington, D.C.: National Academy Press.

Cheatham, N. H., and J. R. Haller. 1975. "An Annotated List of California Habitat Types." Unpublished manuscript, University of California Natural Land and Water Reserves System.

Crumpacker, D. W., S. W. Hodge, D. Friedley, and W. P. Gregg, Jr. 1988. "A Preliminary Assessment of the Status of Major Terrestrial and Wetland Ecosystems on Federal and Indian Lands in the United States." *Conservation Biology* 2:103–115.

Csuti, B. A., J. M. Scott, and J. Estes. 1987. "Looking Beyond Species-Oriented Conservation." *Endangered Species Update* 5(2):4.

Diamond, J. 1986. "The Design of a Nature Reserve System for Indonesian New Guinea." Pp. 485–503 in M. E. Soulé (ed.), *Conservation Biology: The Science of Scarcity and Diversity*. Sunderland, Mass.: Sinauer Associates.

Ehrlich, P. R., and A. H. Ehrlich. 1981. *Extinction*. New York: Random House.

Gilpin, M. E., and M. E. Soulé. 1986. "Minimum Viable Populations:

Processes of Species Extinction." Pp. 19–34 in M. E. Soulé (ed.), *Conservation Biology: The Science of Scarcity and Diversity*. Sunderland, Mass.: Sinauer Associates.

Goodman, D. 1987. "How Do Any Species Persist? Lessons for Conservation Biology." *Conservation Biology* 1:59–62.

Griffith, B., J. M. Scott, J. W. Carpenter, and C. Reed. 1989. "Translocation as a Species Conservation Tool: Status and Strategy." *Science* 245:477–480.

Holland, R. F. 1987. "Is *Quercus lobata* a Rare Plant? Approaches to Conservation of Rare Plant Communities That Lack Rare Plant Species." Pp. 129–132 in T. S. Elias (ed.), *Conservation and Management of Rare and Endangered Plants*. Sacramento: California Native Plant Society.

Huntley, B. J. 1988. "Conserving and Monitoring Biotic Diversity: Some African Examples." Pp. 248–260 in E. O. Wilson (ed.), *Biodiversity*. Washington, D.C.: National Academy Press.

Hutto, R. L., S. Reel, and P. B. Landres. 1987. "A Critical Evaluation of the Species Approach to Biological Conservation." *Endangered Species Update* 4(12):1–4.

Jones and Stokes Associates. 1987. *Sliding Toward Extinction: The State of California's Natural Heritage, 1987*. Commissioned by the California Nature Conservancy for the California Senate Committee on Natural Resources and Wildlife.

Klubnikin, K. 1979. "An Analysis of the Distribution of Park and Preserve Systems Relative to Vegetation Types in California." Master's thesis, California State University, Fullerton.

Margules, C. R., A. O. Nicholls, and R. L. Pressey. 1988. "Selecting Networks of Reserves to Maximize Biological Diversity." *Biological Conservation* 43:63–76.

Mayer, K. E. 1984. "A Review of Selected Remote Sensing and Computer Technologies Applied to Wildlife Habitat Inventories." *California Fish and Game* 70(2):101–112.

Murphy, D. D., and B. A. Wilcox. 1986. "Butterfly Diversity in Natural Habitat Fragments: A Test of the Validity of Vertebrate-Based Management." Pp. 287–292 in J. Verner, M. L. Morrison, and C. J. Ralph (eds.), *Wildlife 2000: Modeling Habitat Relationships of Terrestrial Vertebrates*. Madison: University of Wisconsin Press.

Myers, N. 1979. *The Sinking Ark*. Oxford: Pergamon Press.

Noss, R. F. 1987a. "From Plant Communities to Landscapes in Conservation Inventories: A Look at The Nature Conservancy (USA)." *Biological Conservation* 41:11–37.

Noss, R. F. 1987b. "Protecting Natural Areas in Fragmented Landscapes." *Natural Areas Journal* 7(1):2–13.

Paysen, T. E., J. A. Derby, H. Black, Jr., V. C. Bleich, and J. W. Mincks. 1980. *A Vegetation Classification System Applied to Southern California.* Pacific Southwest Forest and Range Experiment Station, Gen. Tech. Report PSW-45.

Pickett, S.T.A., and J. N. Thompson. 1978. "Patch Dynamics and the Design of Nature Reserves." *Biological Conservation* 13:27–37.

Pyle, R. M. 1982. "Butterfly Eco-geography and Biological Conservation in Washington." *Atala* 8:1–26.

Scott, J. M., B. Csuti, J. D. Jacobi, and J. E. Estes. 1987. "Species Richness: A Geographic Approach to Protecting Future Biological Diversity." *BioScience* 37:782–788.

Smith, K. A. 1988. "Endangered Species Management—A Single Species Approach to the Loss of Biological Diversity?" *Transactions of the Western Section of the Wildlife Society*, vol. 24.

Tangley, L. 1985. "A National Biological Survey." *BioScience* 35:686–690.

Udvardy, M.D.F. 1984. "A Biogeographical Classification System for Terrestrial Environments." Pp. 34–38 in J. A. McNeely and K. R. Miller (eds.), *National Parks, Conservation, and Development.* Washington, D.C.: Smithsonian Institution Press.

U.S. Fish and Wildlife Service (USFWS). 1988. *Endangered Species Technical Bulletin* 8(5).

Wilcove, D. S. 1988. *National Forests: Policies for the Future.* Vol 2: *Protecting Biological Diversity.* Washington, D.C.: Wilderness Society.

Wilson, E. O. 1985. "The Biological Diversity Crisis." *BioScience* 35:700–706.

Wilson, E. O. 1988. "The Current State of Biological Diversity Crisis." Pp. 3–18 in E. O. Wilson (ed.), *Biodiversity.* Washington, D.C.: National Academy Press.

NEEDED: AN ENDANGERED
HUMANITY ACT?

by

ANNE H. EHRLICH AND PAUL R. EHRLICH

A LANDMARK PIECE of conservation legislation, the Endangered Species Act, was recently reauthorized by Congress as it certainly should. Not only has the Endangered Species Act helped directly to preserve biological diversity in the United States, but it has also been an important tool in calling to the public's attention the need for such preservation. And it has kept America a pioneer in conservation legislation, a position it has enjoyed since the establishment of the world's first national park in 1872.

Proud as the United States should be of the act, it is, nonetheless, inadequate in many ways. We therefore would like to explore the directions in which, if it were politically feasible, it could be improved. The comments that follow are based on the assumption that the preservation of biological diversity is, after the prevention of nuclear war, the single most critical task facing humanity. Not only do the other organisms with which we share Earth supply us with enormous aesthetic and economic resources, but they are involved in supplying humanity with an array of free ecosystem services without which civilization simply could not survive. At the present time, biological diversity is in greater jeopardy worldwide than it has been at any time in the last 65 million years. One species, *Homo sapiens*, has entrained an episode of extinctions that promises to exceed even the cataclysm that destroyed the dinosaurs.

The only significant legal tool that the United States has to help counter this ominous trend is the Endangered Species Act. What, then, could be done to enhance its effectiveness?

The first step in strengthening the act would be to broaden the focus in two directions away from "species." In one direction, more attention must be called to the problem of the extirpation of *populations*. Saving the last few members of an endangered species may be important aesthetically and politically—as it is in the case of the California condor, an "umbrella species" whose presence in the wild helps to protect a very large area of habitat containing many other important organisms. But from the standpoints of the practical value of other organisms and the long-term survival of species, it is the numerous genetically differentiated populations *within* species that must be protected. Without genetic diversity within species, not only is the continued existence of the species jeopardized but so is the species' potential for being developed into a commercially valuable domesticated one. And a few individuals of a species (or even a few small populations) are rarely able to participate significantly in the delivery of ecosystem services.

Since distinct populations of most organisms, especially invertebrates and plants, have not even been delimited, they will be virtually impossible to protect one by one. This leads to the other direction that modification of the Endangered Species Act should take—toward an "Endangered Community" or "Endangered Habitat" Act. That is, it should be deemed as a law to maintain the maximum number of relatively undisturbed natural ecosystems of the United States in as pristine a condition as possible. That would automatically promote protection of both populations and ecosystems.

What sorts of devices might be written into a new law to accomplish such ends? Development of relatively undisturbed land should be forbidden unless the developer can demonstrate a critical, long-term *public* benefit. Only in the most unusual circumstances should more virgin land be bulldozed to make way for homes or shopping centers or be brought under the plow, logged, or otherwise seriously disturbed. More than enough territory has already been "developed" to maintain the vast majority of activities that are necessary for the present American population, or even one as large as 300 million or so (if that cannot be avoided). Indeed, the whole notion of a "developer" ought to be converted into that of a "re-developer." The idea that a little bit more land development is harmless must be put to rest once and for all—we have had a lot more than a little bit more, and it has already done severe damage.

Such a law, of course, would be difficult to draft. But the time to start thinking about it was yesterday. Because if there is to be a reasonably equitable halt to further development of virgin lands, grandfathering and subsidies will have to be provided. The law should not place a heavy economic burden on people for actions taken in an era when such development was not only legal but considered desirable. Phasing in a new law would have to be particularly cleverly done to prevent a last-minute rush of development that would further damage the America that we hope to pass on to our grandchildren and great-grandchildren.

But an Endangered Habitat Act should go even further. While making the development of virgin land a privilege reserved only for the most extraordinary circumstances, the law could also encourage steps to make disturbed areas much more hospitable to nonhuman organisms. One need only look into some suburban subdivisions in the American Southwest, for instance, where native vegetation has been retained in landscaping, to see a fascinating diversity of other organisms that can coexist with *Homo sapiens*. A cactus wren or gila woodpecker in one's backyard is a wonderful bonus for not putting in that expensive and thirsty plantation of exotic grasses known as a lawn. Indeed, many ecologists dream of the day when lawns are again a rarity in North America and homes are surrounded primarily by native vegetation. That step alone would be a major stride toward preserving organic diversity in the United States.

An Endangered Habitat Act could also provide for restoration of habitat, even when that habitat is not occupied by a specific threatened organism. When slums are redeveloped to provide decent low-cost housing, for example, areas should be set aside as open space to be planted in native vegetation. Programs should be established to restore stream courses through both cities and farmland to as natural a state as possible; water quality would be enhanced, floods would be less frequent, and important corridors for larger wildlife (and habitat for smaller organisms) would be provided. Obviously, such habitat restoration activities would cost money. But incentives such as tax credits could be applied or, in some situations, penalties for failure to make habitat enhancement part of redevelopment schemes that involve federal funding.

Broadening the focus of the law from species to include populations and communities, and changing the focus from simply preservation to the increase and enhancement of habitat, would be enormous steps in the right direction. But by themselves even they might not be enough. All habitats will remain threatened as long as *Homo sapiens* continue to spew all manner of toxins into the environment, generate acid precipitation, and inject carbon dioxide into the atmosphere, increasing the greenhouse effect.

Clearly it is impossible to protect biological communities from assaults carried in air and water or from the effects of rapid climate change. The most wonderful forest reserve or species-rich lake can be converted into a near biological desert by acid rain. And there is little hope of entire floras and faunas migrating in response to climate change—not only because of the prospective speed of that change but also, perhaps more importantly, because of the barriers created between the remaining fragments of habitat by the works of humanity. Although these problems are transnational, the United States could address these issues and set an example for other nations to follow.

Another way in which the United States could promote the preservation of biotic diversity would be to become the first developed nation to adopt officially the goal of *reducing* its population. This has already been done by the People's Republic of China, but the rich countries whose population growth represents a much more serious threat to the global environment have not yet taken such a step. Instead of showing leadership, the United States has vigorously promoted population growth at home and abroad.

Work that we have done in collaboration with our colleagues Peter Vitousek and Pamela Matson has shown that one of some 5 to 30 million species of the planet, *Homo sapiens*, is now coopting a very large portion of the world's net primary productivity. Net primary productivity (NPP) is the solar energy bound by plants in the process of photosynthesis, except for the fraction used by the plants for their own life processes. The global NPP is the basic food resource for Earth's animals. About 60 percent of that NPP takes place on land, where human impact is by far the heaviest. Humanity coopts about 40 percent of that terrestrial NPP. On this basis alone, human population growth represents a serious threat to Earth's biological diversity. While it would not be legally appropriate for an Endangered Species Act to include provisions to reduce the American birthrate further, few steps would be more beneficial in preserving our life support systems.

We do not, of course, expect these suggestions to be written into law in the near future. But one way or another, the tight connection between our own future and that of our fellow passengers on Spaceship Earth must be made clear to everyone. Perhaps we will know that the message has finally been heard when the Endangered Species Act is combined with population control legislation into an Endangered Humanity Act.

Note

This essay is reprinted with permission from the Spring 1986 issue of the *Amicus Journal*.

INDEX

Achatinella, 182
Acquisition funds, *see* Appropriations for endangered species
Adaptive environmental assessment and management approach, 261
Adaptive management, 158
Agriculture and Wildlife Coexistence Committee, 222
Alabama, 102
Alabama flattened musk turtle, 41
Alligator National Wildlife Refuge, 99
Alligator River National Wildlife River Refuge, 204–205, 206
American alligator, 14, 28, 38, 80, 148, 201
American Farm Bureau Federation, 219
Antioch Dunes evening-primrose, 190
Appropriations for endangered species, 12, 14, 20, 102–104, 134–46, 278
  directions for the future, 143–46
  price of inadequate funding, 140–43, 227–28
Aransas National Wildlife Refuge, 100
Arizona, 52
Army Corps of Engineers, U.S., 175–76

Bald eagle, 27, 31, 38, 39, 64, 145, 148, 182, 201, 214, 228, 249
  anecdote about, 69–70
Bald Eagle Protection Act of 1940, 100
Bay checkerspot butterfly, 89, 186, 188
Bean, Michael J., 37–42, 190–91
Beavers, 233–34
Bell, Griffin, 3
Betts, Richard, 160
Biodiversity, 30, 195, 227–302
  coarse-filter approach to, 268–79
  ecosystem management to maintain, *see* Ecosystem management

ecosystem versus species approach to, 228–30, 249, 283–85
  fine-filter approach to, 268–69
  flagship species, 235
  future for, 238–42
  gap analysis of species richness and vegetative cover, 282–95
  indicator species, 233–34
  keystone species, 233–34
  need for pluralistic strategy, 230–31
  preserving populations, 299
  proposed Act to preserve ecosystems, 195, 277–79, 299–300
  rarity of species, 235–36
  umbrella species, 234–35
  vulnerable species, 235
  wilderness areas as a baseline, 236–38
Biophysical regions, 276–77
Biosphere reserves, 258
Biotechnology, 30
Birds, 185
Bison, 100
Bixby, Kevin, 8
Black–footed ferret, 28–29, 37, 39
  recovery effort for, 148–56, 201
Black pelican, 28
Bolivia, 130
Borax Corporation, 201
Boundary effects, 260
Brown pelican, 38, 80, 214
Buffalo, 98
Bureau of Land Management (BLM), 65, 67, 95, 195, 210, 260, 289
  funding for, 134, 136, 138–39, 143
Bureau of Reclamation, 168–69
Butterflies, 287, 288
  see also specific types of butterflies

Cabinet-Yaak grizzly bears, 205, 206
Cactus wren, 300
Caicco, Steve, 282–94
California, 52, 170, 172, 186, 286
California condor, 28–29, 37, 39, 42, 70, 201, 214, 228, 299
California Department of Fish and Game, 286
California Nature Conservancy, 282–83, 286
California State Water Resource Board, 172
California Supreme Court, 172
Callippe silverspot butterfly, 188
Campbell, Faith, 8, 134–46
Captivity, species surviving in, 29, 201, 203
   captive propagation, 37, 38, 189, 269
   of black-footed ferret, 150, 151, 152, 155
Carolina parakeet, 77
Carson-Truckee Water Conservancy District v. Watt, 176–77
Center for Environmental Conservation, 136
Chimpanzee, 39
CITES, *see* Convention on International Trade in Endangered Species
Citizen suits, 1973 ESA's provision for, 21
Clark, Tim, 147–61
Clean Water Act:
   Section 404, 175–76
      permits, 173–74
   Wallop amendment, 176
Cluster approach to reviewing pesticides, 216–17
Coarse-filter approach to biodiversity, 268–79
   advantages of, 269–70
   explanation of, 268
   improving the, 274–77
      biophysical regions, 276–77
      ecosystems, 274–76
      scale, 277
   limitations of, 270–74
      community dominants, 271–72
      comprehensiveness, 270
      insufficient information, 273–74
      predictable communities, 272–73
      scale, 270–72
   proposed Endangered Ecosystems Act, 277–79
Coggins, George Cameron, 7, 62–71
Colorado, 178, 209

Colorado River, 35, 93
Communities, 285–86
   in coarse-filter strategy, 269–73, 274, 279
Computerized sources of endangered species information, 94
Concho water snake, 41, 92
Conference of the Parties (COP), 123
Consumptive rights, 170, 172
Contra Costa wallflower, 190
Convention on International Trade in Endangered Species (CITES), 14–15, 19, 45, 115, 117–31, 137
   ambiguity of objectives of, 126–27
   enforcement of, 124–25
   evaluation of, 120–21
   growth of CITES appendices, 123–24
   impacts of, 126–27
   lessons from, 127–30
   overview of, 118–19
   practical difficulties of, 121–23
   U.S. responsibilities under, 119–20
Convention on Nature Protection and Wildlife Preservation in the Western Hemisphere of 1940, 20
Cougars, 209
Cracraft, Joel, 54
Cragun, J., 159
Critical habitat designation, 190–91
Crocodiles, 124, 130
Cromack, K., 230
Crumpacker, D. W., 286
Csuti, Blair, 282–94

DDT, 27, 28
Decision analysis, 154–55
Decision seminar model, 157
Defenders of Wildlife, 115, 140, 141
Deforestation, 39, 241
Denison, W., 230
Department of Agriculture, U.S., 134, 138, 219
   funding for, 134, 137
Dingell, John D., 25–30
Dinosaur National Monument, 168
Dunkle, Frank, 108
Dusky seaside sparrow, 28, 39, 227

Eastern timber wolf, 203
Ecosystem conservation, 182, 192, 195, 236–38, 247–63
   defining ecosystems, methods of, 274–76
   through ecosystem management, *see* Ecosystem management

ecosystem level of biological organization, 231
proposed Endangered Ecosystem Act, 277–79
scale of ecosystems, 277
versus species protection, 228–30, 249, 283–85
Ecosystem management, 247–63
adopting plans that integrate geographic scales, 254–57
alternatives to siege mentality, 249–50
capitalizing on constituent support, 261–62, 263
coordinating interagency actions, 257–60, 263
focusing on specific ecosystems and elements of diversity, 251–54
integrating research, development, and application, 260–61, 263
land ethic, 250–51
limitations of protected natural areas, 248
need for encompassing strategy, 248–49
turning the vision into reality, 262–63
Education, environmental, 239
Egypt, 116
Ehrenfeld, David, 5
Ehrlich, Anne, 278, 298–302
Ehrlich, Paul, 278, 298–302
Eldredge, Niles, 54
Elephant, 39, 126
ivory trade, 126–27
El Segundo blue butterfly, 187
Emergency Wetlands Resources Act of 1986, 278
Endangered Community (or Habitat) Act (proposed), 299–302
Endangered Ecosystems Act (proposed), 277–79
Endangered species:
appropriations for, *see* Appropriations for endangered species
failed rescue efforts, 28–29, 39
list of, *see* List of endangered species
state responsibility for, *see* States
see also specific species
Endangered Species Act of 1973, 15–71, 230, 241, 251, 268
as departure from wildlife law, 63–64
federalism and, 98–110
as fine-filter approach, 268–69

funding of, *see* Appropriations for endangered species
history of wildlife intervention, 10–15, 100–101
innovative solutions resulting from, 28, 34, 70, 89, 178
international implications of, *see* International wildlife conservation
key provisions of, 15–21
misuse of, 35–36
1978 amendments to, 4, 16, 17, 45, 79, 187, 191
1982 amendments to, 16, 18, 45, 47, 178, 204
1988 amendments to, 17, 19, 47, 50, 51, 84, 103, 107, 108, 219
passage of, 5, 15–16
pesticide regulation and, 214–23
predator conservation and, 199–211
reflections on, 70–71
Section 5 (habitat acquisition), 15, 20, 170
Section 6, 20, 99, 101–104, 141, 145
Section 7 (interagency consultation), 17–18, 34, 40, 50, 51, 86–96, 115–16, 141, 190–91, 210, 215–18
Committee to review exceptions to, 17, 27, 32, 45, 57, 62, 87
Section 8 (international provisions), 19–20, 114, 116–17
Section 9 (restrictions on trade), 19
species approach of, weaknesses of, 229, 249, 283–85
taking provision, 15, 18–19, 47, 49–50, 101, 104–105
western water rights and, 167–79
see also Biodiversity; Ecosystem conservation
Endangered Species Act Reauthorization Coalition (ESARC), 137, 140
Endangered Species Information System, 194
Endangered Species Preservation Act of 1966, 5, 101
intent of, 6–7, 78
provisions of, 12–13, 78
Endangered Species Preservation Act of 1969, 5, 101, 114, 195
provisions of, 13–15, 78
Endangered Species Update, 5
Environmental impact statement (EIS), 168–69

Environmental Protection Agency
(EPA), pesticides and, 215–22
Ernst, John P., 98–110
Estes, J. E., 282–94
Experimental populations, 18, 19, 204
Extinction of species, 4, 28–29, 39, 227
rate of, 4–5, 29, 283

Federal agencies, 38, 40
interagency cooperation, *see*
Interagency cooperation
major land management agencies,
64–65
see also specific agencies
Federalism and ESA, 98–110
early state control of wildlife, 99–100
history of federal intervention, 100–
101
issues and prospects, 109–10
Section 6 cooperative agreements,
101–104
funding for, 102–104
taking provisions and controversy,
104–109
Federal Power Act, 177
Federal public lands, 64–68
ESA lawsuits concerning, 65–68
predator conservation of, 208, 209
Federal Register, 12, 183
Fish and Wildlife Coordination Act of
1934, 11
Fish and Wildlife Service (FWS), 16, 17,
18, 31, 67, 143, 173–74, 175
allocation of funds, 144
black-footed ferret recovery plan
and, 149, 151
consultation process and, 87–94, 116,
141, 173, 215–17
training for, 93, 94
decision making by, 65, 66, 178
definitions set by, 18
directors of, 108, 109
funding for, 134, 135–36, 143, 228
international implications of ESA
and, 115–17
invertebrate conservation and, 183–
93
listing and, 80, 81, 149, 228, 269, 282
political pressures on, 41, 42, 65, 92,
108
predator conservation and, 202, 203,
204, 208
recovery implementation reports,
141
taking by, 106

Flagship species, 235
Florida, 52
Florida panther, 235
Forest Service, 65, 66, 67, 95, 195, 202,
210, 232, 233, 236, 258, 260
funding for, 134, 136, 138, 143
Franklin, J. F., 230
Fund for Animals v. Andrus, 106
Funding, *see* Appropriations for
endangered species
Fur industry, 14

Gap analysis of species richness and
vegetative cover, 282–94
biodiversity outside reserves, 292–94
explanation of the concept, 285–88
field verification of sites, 292
materials and methods for, 288–92
model program, 294
Geer v. Connecticut, 100
General Accounting Office (GAO), 90
Genetic level of biological
organization, 230, 231
Geographic information systems (GIS),
289–94
Geographic scales for ecosystem
management, 254–57
Giant panda, 39
Gila woodpecker, 300
Gilbert, L. E., 233
Glen Canyon Dam, 168–69
Goal displacement, 152
Golden lion tamarin, 235
Gopher tortoise, 234
Grayrocks Dam, 93, 173–74
Gray bats, 214
Gray wolf, 98–99, 105–10, 182, 201,
249
reintroduction of, 203
Great auk, 77
Green River, 168
Greenwalt, Lynn A., 31–36
Gregg, W. P., Jr., 286
Grimm, E. C., 272
Grizzly bear, 33, 66, 67, 99, 105–10,
182, 186, 201, 202, 205–209, 249,
255, 262

Habitat acquisition, 240, 269
funding of, 12, 14, 20, 141–43, 278
history of, 11, 12, 14
for invertebrate recovery, 190
1973 ESA and amendments to, 20
Habitat protection, 88, 167, 229, 240

Endangered Habitat Act (proposed), 299–302
see also Biodiversity; Ecosystem management
Hakalau Forest National Wildlife Refuge, 142–43, 145
Haleakala National Park, 139
Harris, L. D., 260
Harvey, Ann, 147–61
Hawaii, 18, 52, 139, 142
Hawaii Volcanos National Park, 139
Heath hen, 100
History of federal wildlife management legislation, 10–15, 100–101
Hodge, S. W., 286
Holling, C. S., 261
Howell, Ed, 98
Hummingbird species, 185
Humpbacked chub, 168, 169, 178
Hunter, Malcolm L., Jr., 266–79
Hunting, *see* Taking of species

Idaho, 107, 202, 294
Incidental takings, 18, 19
India, 116
Indicator species, 192, 232–33, 272
Information systems, 94
  intelligence failures in recovery programs, 153–56
Insects, *see* Invertebrate conservation
Interagency Agreement on Spotted Owl Management, 258
Interagency cooperation:
  for ecosystem management, 254, 257–60, 263
  implementation record, 89–91
  improving the process, 92–96
  international application of, 115–16
  1966 Act on, 13
  1973 ESA and amendments on, 17–18, 86–96
  problems in implementation, 91–92
  success of, 87–89
Interagency Grizzly Bear Management Committee (IGBC), 202, 254, 258
Interior Department, U.S., 11, 14, 16
  Bureau of Sport Fisheries and Wildlife, Committee on Rare and Endangered Species, 11–12
  see also Secretary of Interior
International Union for the Conservation of Nature and Natural Resources, 45
International wildlife conservation, 19–20

conferences on, *see* Convention on International Trade in Endangered Species
international implementation of ESA, 114–31
  CITES and, *see* Convention on International Trade in Endangered Species
  Section 7 and, 115–16
  Section 8 and 8A(e), 116–17
  1969 Act on, 14, 114–15
Interstate commerce:
  Lacey Act and, 10, 100
  1966 Act and, 13
  1969 Act on, 13–14
Invertebrate conservation, 181–95, 228
  critical habitat designation, 190–92
  current state of, 181–82
  future prospects of, 192–95
  FWS staff biases, 183–84
  invertebrate subspecies, 187–88
  priority, 185–87
  public perceptions and, 184–85
  recovery plans, 188–90
Invertebrate distribution maps, 288–89
Island biogeographic theory, 257
IUCN, 286
IUCN Captive Breeding Specialist Group (CBSG), 157
Ivory-billed woodpecker, 80

Jacobson, George L., Jr., 272
Johnson, K., 260
Juday, G., 230

Kcauhou-Kilauea National Wildlife Refuge, 142
Kellert, Steven, 184
Keystone species, 192, 233–34
Kohm, Kathryn A., 3–21
Kosloff, Laura H., 114–31

Labrador duck, 77
Lacey Act of 1900, 10–11, 100
Lahontan Valley, 179
Land acquisition, *see* Habitat acquisition
Land and Water Conservation Fund, 12, 141
Land and Water Conservation Fund Act of 1965, 20
Land Between the Lakes region, 204
Land ethic, 250–51
Landownership and management status maps, 289

Landres, P. B., 254
Landscape as geographic scale, 255, 256
Lange's metalmark, 190
Last Extinction, The (Ehrenfeld), 5
Lawrence, Kansas, 68–69
Lawsuits:
  citizen, 21
  concerning federal public land, 65–68
  see also specific cases
Leahy, Patrick, 4
Leopards, 126
Leopold, Aldo, 84, 227, 228, 238, 241, 250
List of endangered species, 71, 77–84, 149
  assessment of the past, 82–83
  backlog in, 17, 29, 39, 79, 81, 140, 227–28, 247, 269, 282
  current listing status, 79–80
  delistings, 80
  effects of listing, 80–81
  endangered versus threatened categories, 16
  first ("Redbook"), 12, 78, 149
  invertebrates on, 181–82
  under 1966 Act, 12–13, 78
  under 1969 Act, 14, 78
  under 1973 ESA and amendments, 16, 79
  number of species on, 6–7, 17, 29, 39, 79, 83, 247, 269, 282
  recovery plans for, *see* Recovery plans
  suggestions for, 145
Lobbying efforts, 14
  for appropriations, 145
  pesticides and, 219
  against predators, 200
Locke, John, 53
Longleaf pine ecosystem, 234, 237
Louisiana, 102
Lujan, Manuel, 169

MacBryde, Bruce, 48
McClure, James, 108, 110
Mckee, A., 230
Maine, 276
Management indicator species, 232–33
Maps, 288–89
Marine Mammal Protection Act of 1972, 100, 101
Marlenee, Ron, 108
Martin v. Waddell, 99
Maser, C., 230

Maslyn, Mark, 219
Matson, Pamela, 301
Meadows, 233–34
Mealey, S. P., 260
Metapopulations, 191
Mexican wolf, 201, 203, 205–206, 209, 210
Meyers, Norman, 5
Michigan, 203
Migratory Bird Act of 1913, 11
Migratory Bird Conservation Act, 11
Migratory Bird Hunting Stamp Act of 1934, 11
Migratory Bird Treaty of 1918, 11, 100
Minimum critical ecosystem size, 260
Minimum dynamic area, 260
Minnesota, 208, 214
Minnesota Wolf case, 67, 208
Monarch butterflies, 255
Mono Lake brine shrimp, 188
Montana, 107, 108, 202, 208–209
Montana Department of Fish, Wildlife, and Parks (MDFWP), 106–107
Moore, Jack, 217
Mountain States Legal Foundation, 219
Mount Graham red squirrel, 41, 241
Multiple-use lands, 248
Muriqui, 235
Murphy, Dennis D., 181–95
Myers, Norman, 45–46, 283

National Audubon, 172
National Ecosystems Act (proposed), 195
National Environmental Policy Act (NEPA) 168, 177, 251
National Forest Management Act of 1976, 232, 251
National forests, 256
National Marine Fisheries Service (NMFS), 16, 17, 18, 80, 87
  allocation of funds, 144
  funding for, 134, 136–37, 143
National Oceanic and Atmospheric Administration, 42
National parks, 229, 256, 257, 298
National Park Service (NPS), 65, 195, 202, 203, 260
  funding for, 139
National Wilderness Preservation System, 238
National Wildlife Refuge System, 65
Native Americans, water rights of, 171
Natural values versus noneconomic values, 44–46

Nature Conservancy, The, 38, 236, 268, 289
  California, 282–83, 286
  Nebraska, 170, 173–74
  Nebraska State Department of Water Resources, 174
Nebraska v. REA, 67, 173–74
Net primary productivity (NPP), 301–302
Newmark, W. D., 257
Nixon, Richard, 15, 38
Noneconomic values versus natural values, 44–46
North American Wetlands Conservation Act of 1989, 278
North American Wild Sheep, 66
North Carolina, 102
North Platte River, 88, 95, 175
Northern crawfish frog, 68–69
Northern spotted owl, 183
Northwest Power Act, 177
Noss, Reed F., 227–42, 286, 292, 293
Notification processes, 94

Office of Management and Budget, 138
Ohio, 103
Oregon, 52
Organizational structures for recovery programs, 152–53, 154
  improving, 158–59
Owens, Wayne, 108

Paine, R. T., 233
Pakistan, 116
Palila v. Hawaii Department of Land and Natural Resources, 18, 52, 104
Palos Verdes blue butterfly, 28, 39, 227
Parks as island faunas, 260
Passenger pigeon, 77
Pelican Island National Wildlife Refuge, 11
Penalties, 20–21, 125
Peregrine falcon, 28, 38, 145, 148, 255, 261
Pesticide regulation and ESA, 214–23
  compliance problems, 217–20
  consultation history, 216–17
  evaluation of, 220–22
  Section 7 responsibility, 215–16
Pickett, S. T. A., 292
Pittman-Robertson Act, 11
Plants, 47–48, 52, 54, 78, 228
  gap analysis of species richness and vegetative cover, 282–95

1988 amendments to ESA and, 19, 47, 51, 84
  taking of, 50, 51
Platte River, 173, 174
Pocket gophers, 233–34
Political pressures, 34, 40–42, 65, 92, 228
  gaining support for ecosystem approach to conserving endangered species, 261–62, 263
  pesticides and, 219, 222
  predator conservation and, 200, 202, 203, 204
  wolf reintroduction and, 108
Population reduction, 301, 302
Populations, preserving, 299
Power goals, 152
Predator conservation, 199–211
  attitudes toward, 200, 202
  complications of, 200
  federal resources for, 208–209
  history of treatment of predators, 199
  lessons learned about, 209–11
  negotiated recovery actions, 204–206
  new phase in, 202–204
  the record on, 200–202
  research support, 201
  zone management, 206–207
Pregerson, Judge, 177
Prior appropriation, law of, 170–71
Priority setting, 145, 263
  invertebrates and, 185–87
Private property, flora and fauna on, 46–54
  question of gain or loss, 51–54
  taking of, 48–49
Probabilistic models, 158
Pronghorn antelope, 186
Public water rights, 171–72
Puerto Rico, 52
Pyramid Lake, 176

Quino checkerspot butterfly, 184

Reagan, Ronald, and Reagan administration, 102, 108, 116, 136, 140, 144, 202
Recovery plans, 29, 79, 80, 140–41, 227, 283
  implementation problems of, 147–56
    complexity of cooperation, 151
    goal displacement, 152
    intelligence failures, 153–56
    organizational structures, 152–53, 154

Recovery plans (*continued*)
  improvements in, 156–60
    of the organization, 158–59
    of the people, 159–60
    of the process, 156–58
  invertebrates and, 188–90
  predators and, 204–206
Red-cockaded woodpecker, 67, 234, 237, 255
Red-naped sapsuckers, 234
Red wolf, 102, 201, 202, 204–205, 206, 214
Reffalt, William, 6, 77–84, 142
Refuge management, 142–43, 145
Region as geographic scale, 254–55, 259
Regulatory water right, 170, 171, 173
Research into ecosystems, integrative, 260–61, 263
Research natural areas (RNAs), 237
Rhinoceros, 39, 126
Rhododendron chapmanii, 48, 55, 56
Riparian rights, 170
Riverside Irrigation District v. Andrews, 67, 174–76
Rolston, Holmes, III, 43–61
Rosy periwinkle, 26
Rural Electrification Administration (REA), 173

Sacramento-San Joaquin Delta, 172
Salwasser, Hal, 247–63
Sandhill cranes, 27
San Diego mesa mint, 48
San Nicholas Island, 205, 206–207
Scott, J. Michael, 8, 194–95, 282–94
Sea otter, 202, 205, 206–207
Sea turtles, 28, 39, 126, 214
Secretary of Agriculture, 10–11
  cooperative agreements with states, 20
  habitat acquisition and, 20
Secretary of Interior, 176–77, 178
  acquisition authority, 12
  CITES and, 14–15, 114–15
  cooperative agreements with states, 20, 102
  foreign endangered species programs and, 20
  habitat acquisition and, 20
  list of species threatened with worldwide extinction and, 14, 18
Sedell, J., 230
Senate, U.S., Environment and Public Works Committee, 108
Serfis, Jim, 7, 214–23
Shasta crayfish, 187

Sierra Club, 18, 168
Sierra Club v. Clark, 106
Simpson, Alan, 107–108
Simpson, G. G., 54
Smith, Kent, 282–94
Snail darter, 3, 16, 17, 26–27, 32–33, 40, 214
Socorro isopod, 31
South Platte River, 175
Species:
  concern for individuals versus, 54–58
  ecosystem protection versus protection of, 228–30, 249, 283–85
  extinction of, *see* Extinction of species
  flagship, 235
  identified versus unidentified, 266–67
  indicator, 232–33, 272
  keystone, 233–34
  popularity of, 184–85, 249
  rare, 235–36
  umbrella, 192, 234–35, 299
  vulnerable, 235
Spotted owl, 67, 236–37, 249, 255, 256
Stacey Dam, 41, 92
Stampede Dam case, 67
Stampede Reservoir, 177
Stand/site as geographic scale, 255, 256
States:
  conflicts with federal government over taking, 104–109, 208
  cooperative agreements with federal government, 20, 101–104
    funding for, 102–104, 141
  endangered species legislation, 38
    private property and, 47
  federal conflict over water rights, *see* Water rights and ESA, western
  historical responsibility for wildlife of, 99–100, 208, 210
  interstate commerce, *see* Interstate commerce
  land-use restrictions, 49
  property rights law of, 50
Steel Shot case, 67
Stephen's kangaroo rat, 183
Stillwater Wildlife Management Area, 179
Strychnine case, 67
Sunrise Hunting case, 67
Supreme Court, U.S., 11, 40, 43, 52, 174
  Geer v. Connecticut, 100
  Martin v. Wadell, 99
  TVA v. Hill, 3, 16, 17, 26–27, 32, 88, 169–70
Swanson, F., 230

Taking of private property, 48–49
Taking of species:
  conflicts between state and federal
    governments, 104–109
  definitions of take, 50, 104–105
  exceptions to ESA, 18–19
  history of federal regulation of, 11
  1966 Act on, 13
  1973 ESA and amendments on, 15,
    18–19, 47, 49–50, 101, 104–105,
    178
  of predators, 208–209
  on private property, 46–54
Tarlock, A. Dan, 7, 167–79
Task goals, 152
Tellico Dam, 3, 26–27, 32, 40, 88, 169,
  214
Tennessee-Tombigbee Waterway, 41
Tennessee Valley Authority (TVA), 3
  TVA v. Hill, 3, 16, 17, 26–27, 32–33,
    88, 169–70
Texas, 52, 170, 203, 209
"Thinking Like a Mountain," 84
Thomas, J. W., 254, 260
Thompson, J. N., 292
Threatened species:
  versus endangered species, 16
  taking of, 106
Timber wolf, 33
Tombigbee River freshwater mussels,
  41
Trade restrictions, 114
  CITES and, *see* Convention on
    International Trade in
    Endangered Species
  interstate commerce, *see* Interstate
    commerce
  1973 ESA and amendments on, 19
Trexler, Mark C., 114–31
Trumpeter swans, 255
Tule elk, 186
Turner, John, 109
Two-Forks Dam, 41

Udall, Stewart, 12
Umbrella species, 192, 234–35, 299
UNESCO, 286
University of Michigan, School of
  Natural Resources, 5
Upper Colorado River, 88, 93, 95
Utah, 52, 178
Utilitarianism in wildlife law, 63

Vegetative classification, 285–86
Vegetative maps, 288, 289
Verner, J., 254
Vertebrate distribution maps, 288–89

Vitousek, Peter, 301
Vulnerable species, 235

Walters, C., 261
Washington, 170, 202
Washoe Project Act, 177
Water rights and ESA, western, 167–79
  consumptive rights, 170, 172
  federal/state water conflicts, 170
  finding middle ground, 178–79
  Glen Canyon Dam controversy, 168–
    69
  law of prior appropriation, 170–71
  major court cases, 173–77
  public rights, 171–72
  regulatory water rights, 170, 171, 173
  riparian rights, 170
Watershed as geographic scale, 255, 256
Webb, T., III, 272
Weinberg, David, 152
Western Area Power Administration,
  169
Western Hemisphere Convention, 115,
  117
Western water rights, *see* Water rights
  and ESA, western
West Indian manatee, 89
Wetlands, 233–34
White Sands Missile Range, 203
Whooping cranes, 27, 38, 173–74, 175
Wildcat Reservoir, 175
Wilderness:
  as baseline for managed ecosystems,
    236–38
  designation, 238
Wild Free-Roaming Horses and Burros
  Act of 1971, 100
Wildlife Refuges System, 195
Windy Gap strategy, 93
Wilson, E. O., 181
Wiregrass, 234
Wolves, 255
  see also specific types of wolves
Wyoming, 107, 108, 153, 155, 157, 158,
  202, 203, 209
Wyoming Game and Fish Department,
  149, 151, 152

Xerces Society, 185

Yaffee, Steven, 7, 8, 86–96
Yellowstone National Park, 11, 98, 105,
  107, 108, 109, 203, 204, 209
Yellowstone Park Protection Act of
  1894, 98

Zone management, 206–207

ABOUT THE CONTRIBUTORS

MICHAEL J. BEAN is chairman of the Environmental Defense Fund's Wildlife Program and author of *The Evolution of National Wildlife Law*.

KEVIN BIXBY is a freelance writer specializing in wildlife conservation issues.

STEVE CAICCO is a plant geologist formerly with the Idaho Department of Fish and Game's Heritage Program.

FAITH CAMPBELL is a senior research associate for the Natural Resources Defense Council.

TIM CLARK is president of the Northern Rockies Conservation Cooperative.

GEORGE CAMERON COGGINS is the Frank Edwards Tyler Professor of Law at the University of Kansas.

BLAIR CSUTI is a research associate with the University of Idaho.

CONGRESSMAN JOHN D. DINGELL has served as chairman of the House Energy and Commerce Committee since 1981 and is widely recognized as the father of some of the nation's most important environmental legislation, including the ESA.

ANNE H. EHRLICH AND PAUL R. EHRLICH are ecologists at Stanford University.

JOHN P. ERNST is director of legislative affairs for the National Wildlife Federation.

J. E. ESTES is a professor of geography and director of the Geography Remote Sensing Unit at the University of California, Santa Barbara.

LYNN A. GREENWALT, director of the Fish and Wildlife Service from 1973 to 1981, is now vice president for international affairs at the National Wildlife Federation.

ANN HARVEY is a research associate at the Northern Rockies Conservation Cooperative.

MALCOLM L. HUNTER, JR., is a professor in the Wildlife Department at the University of Maine where he focuses his research on biodiversity in forest ecosystems and developing countries.

KATHRYN A. KOHM is an editor and writer specializing in biodiversity conservation and other natural resource issues.

LAURA H. KOSLOFF is a trial attorney with the U.S. Department of Justice, Environment and Natural Resources Division.

DENNIS MURPHY is director of the Center for Conservation Biology at Stanford University.

REED F. NOSS is interested in a broad range of conservation strategies and has published many scientific papers on ecological and conservation topics.

WILLIAM REFFALT spent twenty-three years with the Fish and Wildlife Service and is currently the program director for National Wildlife Refuge System issues at the Wilderness Society.

HOLMES ROLSTON III, a professor of philosophy at Colorado State University, is associate editor of *The Journal of Environmental Ethics* and the author of *Philosophy Gone Wild*.

HAL SALWASSER is the director of New Perspectives for the U.S. Forest Service and has served as deputy director of wildlife and fisheries for the Forest Service.

J. MICHAEL SCOTT is a U.S. Fish and Wildlife Service employee and leader of the Cooperative Fish and Wildlife Research Unit at the University of Idaho.

JIM SERFIS, currently with the EPA, previously worked for the Center for Environmental Education where he conducted an independent review of the EPA's compliance with the Endangered Species Act in relation to pesticide registration.

KENT SMITH is coordinator of the Nongame Bird and Mammal Program for the California Department of Fish and Game.

A. DAN TARLOCK is a professor of law at Chicago-Kent College of Law, Illinois Institute of Technology.

MARK C. TREXLER is an associate for the World Resources Institute Program in Climate, Energy, and Pollution in Washington, D.C.

STEVEN L. YAFFEE is a professor in the School of Natural Resources at the University of Michigan and author of *Prohibitive Policy: Implementing the Federal Endangered Species Act*.

ALSO AVAILABLE FROM ISLAND PRESS

Ancient Forests of the Pacific Northwest
By Elliott A. Norse

Better Trout Habitat: A Guide to Stream Restoration and Management
By Christopher J. Hunter

The Challenge of Global Warming
Edited by Dean Edwin Abrahamson

Coastal Alert: Ecosystems, Energy, and Offshore Oil Drilling
By Dwight Holing

The Complete Guide to Environmental Careers
The CEIP Fund

*Creating Successful Communities: A Guidebook for Growth
Management Strategies*
By Michael A. Mantell, Stephen F. Harper, and Luther Propst

Crossroads: Environmental Priorities for the Future
Edited by Peter Borrelli

Economics of Protected Areas
By John A. Dixon and Paul B. Sherman

*Environmental Disputes: Community Involvement in Conflict
Resolution*
By James E. Crowfoot and Julia M. Wondolleck

The Forest and the Trees: A Guide to Excellent Forestry
By Gordon Robinson

*Forests and Forestry in China: Changing Patterns of Resource
Development*
By S.D. Richardson

*Fighting Toxics: A Manual for Protecting Your Family, Community, and
Workplace*
By Gary Cohen and John O'Connor

317

Hazardous Waste from Small Quantity Generators
By Seymour I. Schwartz and Wendy B. Pratt

In Praise of Nature
Edited and with essays by Stephanie Mills

The Living Ocean: Understanding and Protecting Marine Biodiversity
By Boyce Thorne-Miller and John Catena

Natural Resources for the 21st Century
Edited by R. Neil Sampson and Dwight Hair

The New York Environment Book
By Eric A. Goldstein and Mark A. Izeman

Permaculture: A Practical Guide for a Sustainable Future
By Bill Mollison

Plastics: America's Packaging Dilemma
By Nancy A. Wolf and Ellen D. Feldman

Race to Save the Tropics: Ecology and Economics for a Sustainable Future
Edited by Robert Goodland

Recycling and Incineration: Evaluating the Choices
By Richard A. Denison and John Ruston

The Rising Tide: Global Warming and World Sea Levels
By Lynne T. Edgerton

Rivers at Risk: The Concerned Citizen's Guide to Hydropower
By John D. Echeverria, Pope Barrow, and Richard Roos-Collins

Saving the Tropical Forests
By Judith Gradwohl and Russell Greenberg

Shading Our Cities: A Resource Guide for Urban and Community Forests
Edited by Gary Moll and Sara Ebenreck

Wetland Creation and Restoration: The Status of the Science
Edited by Mary E. Kentula and Jon A. Kusler

Wildlife and Habitats in Managed Landscapes
Edited by Jon E. Rodiek and Eric Bolen

Wildlife of the Florida Keys: A Natural History
By James D. Lazell, Jr.

For a complete catalog of Island Press publications, please write:
Island Press
Box 7
Covelo, CA 95428
or call: 1-800-828-1302